Billions of dinosaur tracks have been found in recent years and through careful examination of these prehistoric clues, dinosaur trackers have discovered much about how and where dinosaurs lived. *Tracking Dinosaurs* is the first nontechnical, popular science book on dinosaur footprints and what they reveal about dinosaurs and their habitats. This book deals with the landslide of new information that accumulated in recent decades, demonstrating that fossil footprints are neither rare nor insignificant as previously supposed.

A complete guide to dinosaur tracking, the book begins with a discussion of the meaning of tracks, how tracks provide information about dinosaur locomotion, behavior, ecology and environmental impact. A detailed description of dinosaur track locations is included. Popular myths and misconceptions are reviewed. Did brontosaurs really swim? Did dinosaurs travel in structured herds? These and other questions are discussed in an easy and accessible writing style. Numerous illustrations, including eight pages of color photographs, supplement the text. Everybody who has ever been fascinated by dinosaurs will find this book of interest, as will those with a general interest in science and natural history.

Tracking Dinosaurs

Tracking Dinosaurs
A New Look at an Ancient World

MARTIN LOCKLEY
University of Colorado at Denver

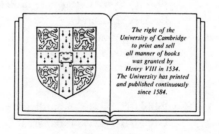

The right of the
University of Cambridge
to print and sell
all manner of books
was granted by
Henry VIII in 1534.
The University has printed
and published continuously
since 1584.

CAMBRIDGE UNIVERSITY PRESS
Cambridge
New York   Port Chester   Melbourne   Sydney

Published by the Press Syndicate of the University of Cambridge
The Pitt Building, Trumpington Street, Cambridge CB2 1RP
40 West 20th Street, New York, NY 10011, USA
10 Stamford Road, Oakleigh, Melbourne 3166, Australia

© Cambridge University Press 1991

First published 1991

Printed in the United States of America

Library of Congress Cataloging-in-Publication Data
Lockley, M. G.
Tracking dinosaurs : a new look at an ancient world / Martin
Lockley.
p.   cm.
Includes bibliographical references and index.
1. Dinosaur tracks. I. Title.
QE862.D5L575   1991
567.9' – dc20   91-2409
                                                          CIP

British Library Cataloguing in Publication Data
Lockley, Martin G.
Tracking dinosaurs.
1. Fossil reptiles
I. Title
567.91

ISBN 0-521-39463-5 hardback
ISBN 0-521-42598-0 paperback

Illustrations copyright © Martin Lockley

To Linda Dale

Contents

Preface

The journey of a thousand miles begins with a single step.
— Lao Tse

In recent years vertebrate paleontology, and the study of dinosaurs in particular, has enjoyed increasing popularity both in the scientific community and among lay persons. This so-called dinosaur renaissance has led to renewed interest in vertebrate ichnology, the study of tracks. This in turn has resulted in an unprecedented spate of footprint discovery and research. Such rapid progress inevitably leads to fresh knowledge and opportunities to view dinosaurs from new perspectives. Whereas tracks were once considered enigmatic oddities or curiosities, they are now recognized as scientifically important data and command increasing attention. The main objective of this book is to convey to the lay reader the substantial contribution that tracks make to our understanding of the Age of Dinosaurs. The professional reader should also benefit from the book because it deals with many new discoveries and synthesizes much of what is now known.

Dinosaur tracks are an unusual and fascinating phenomenon. More than most other evidence from the Age of Dinosaurs, they tell us much about the day-to-day activity of a multitude of extinct species in many different habitats. Footprints come in all shapes and sizes, from birdlike, chicken-size tracks to giant brontosaur footprints, three times larger than those of elephants. They are known to occur in sedimentary rocks from all continents except the Antarctic and range in age from at least 230 million years to 65 million years. Some sites exhibit only a handful of footprints, whereas others yield thousands, even millions, of tracks in areas that extend for thousands of square kilometers.

Footprints are what paleontologists call trace fossils, that is, tracks and traces left by once living animals. They are distinct from body fossils, the actual skeletons and bones, which only enter the fossil record after the animals have died. This distinction is significant in understanding where to look for dinosaur tracks and how to interpret them.

With our present knowledge we can view dinosaur tracks as useful in providing two major categories of information. The first sort is paleobiological, pertaining to the behavior of living dinosaurs. The second type is paleoenvironmental, relating to information that tracks reveal about ancient dinosaur habitats. This book reviews what is currently known about dinosaur tracks, with equal emphasis on what tracks tell us about the dinosaurs themselves and the habitat or environment in which they lived. The mainly biological analyses and interpretation are discussed in the first part of the book, and the paleoenvironmental analysis and discussion are reserved for the second.

Because the science of dinosaur tracking is progressing so rapidly, many long-standing myths are being dispelled or at least discredited. In the concluding chapters we, first, review some of these myths and then look at the science from a historical perspective. Much of what we know today is based on new evidence that has not been fully synthesized into models or hypotheses subjected to the test of time. Nevertheless, such strides have been made that we feel confident that the science of ichnology will never again be considered a trivial pursuit outside the mainstream of paleontology and earth science. Sedimentary strata from the Age of Dinosaurs are replete with footprints. The volume of available evidence is quite staggering, and the implications of the evidence for an improved understanding of dinosaurs are substantial and far reaching.

On this note I turn to my personal perspective, developed after a decade of tracking dinosaurs. As a field-oriented geologist rather than a theoretician, my understanding of dinosaur tracks stems largely from studying them at first hand, at literally hundreds of tracksites in North America, East Asia, and Europe. I have been fortunate enough to be part of a research team, the University of Colorado at Denver Dinosaur Trackers Research Group, that has documented a large number of sites, many new to science. We have benefited, I believe, from the staggering volume and variety of evidence available at our doorstep. We now know of several hundred sites in our region. We find them or have them reported to us at the rate of about fifty a year, or one every week.

The Colorado Plateau Region is without doubt an unparalleled natural, outdoor laboratory for the Age of Dinosaurs. In such an ideal setting one sees tracks of all types and in all states of preservation. One learns to distinguish what is common or rare, normal or unexpected. One can test and compare observations from site to site, analyze patterns and trends, and view tracks from a broad perspective. The sheer volume of material often makes us hesitant to read too much into apparently unusual evidence at a single site. One sometimes learns by trial and error, formulating hypotheses and multiple hypotheses that can be tested, substantiated, or refuted by an ever-expanding volume of new evidence. Sometimes the approach appears haphazard or intuitive, but this is just as much a part of science as the more rigorous, measured approach we try to adopt. Geology and paleontology are often inexact sciences, and dinosaur tracking is sometimes no exception to that rule.

Having issued this disclaimer, it is fair to say that dinosaur tracking is now a respectably established new branch of paleontology and earth science, much more of an exact science than it was a decade ago, and light years ahead of its state of exactness in the many less memorable decades of former generations. The 1980s were an exciting decade, a time that brought, as Dan Chure remarked, an unprecedented spate of research on all aspects of dinosaur tracking. Being a part of this has been exhilarating. We have witnessed the first International Conference on Dinosaur Tracks, which spawned the first book, *Dinosaur Tracks and Traces*, followed by this volume and one other.

This progress could not have been made without the contributions of several dozen researchers. It has been my pleasure and privilege to work with many of the trackers mentioned in this book, to learn from their discoveries and experiences, and to share insights, field observations, and the thrill of science in the wilderness and the Age of Dinosaurs. Some of my closest friends, colleagues, and assistants include present and former members of the Dinosaur Trackers Research Group: Emily Bray, Kelly Conrad, Farley Fleming, James Halfpenny, Karen Houck, Linda Dale Jennings, Mark Jones, Todd McMahon, Eddy Mueller, Michael Parrish, and Nancy Prince. In addition, I have worked closely in the field with Andrew Cohen, University of Arizona; Seong-Kyu Lim and Seong-Young Yang, Kyungpook National University, South Korea; Masaki Matsukawa, Ehime University, Japan; Joaquin Moratalla, Universidad Autonoma Madrid, Spain; Jeffrey Pittman, University of Texas; Bill Hawes and Paula Ott, Grand Junction, Colorado; and Fran and Turby Barnes, Moab, Utah.

Other colleagues and co-authors who have been an inspiration and consistently supportive of the cause of dinosaur tracking have been Kevin Padian, University of California; Paul Olsen, Columbia University; Donald Baird, Princeton University; Dan Chure, Dinosaur National Monument; Georges Demathieu, Dijon University, France; James Farlow, Indiana University; Sue Ann Bilbey and Aldon Hamblin, Utah Field House, Vernal; Phil Currie, Tyrell Museum of Paleontology; and Giuseppe Leonardi, Istituto Cavanis, Venice, Italy. Among these illustrious persons I particularly wish to thank Kevin Padian for his support and encouragement and Don Baird and another anonymous reviewer for reading this book when it was in rough draft form. Father Giuseppe Leonardi has also been a constant source of inspiration, not just in his tracking, which has spanned the entire continent of South America, but also as something of a spiritual mentor.

Others who have helped in various ways include Harley Armstrong, Lani Duke, and Michael Perry of the Museum of Western Colorado (Dinosaur Valley), Karl Hirsh, Chris Schenk, Andrew Rindsburg, Deborah Adelsperger, Bennett Young, and Kenneth Carpenter, all associated at some time with the University of Colorado; Julie Howard of the Bureau of Land Management; James Madsen, former Utah State Paleontologist; and David Gillette, current Utah State Paleontologist and co-editor of *Dinosaur Tracks and Traces*. Many of the friends of Dinosaur Ridge, including Joe Tempel, Dick Scott, Sally White, John Dolson, Karen Hester, Craig Munsart, and others, have also helped the cause of dinosaur tracks in various ways. Among the many filmmakers that have taken an interest in dinosaur tracks, I particularly thank Sir David Attenborough, the BBC, and the New Jersey Network for thoughtful documentary coverage. My thanks go out to all these people and to the many others who have accompanied and assisted us in the field and lab over the years.

I also extend special thanks to the artists who have furnished illustrations for this book. These include Doug Henderson, Michel Henderson, John Sibbick, Donna Sloan, and Eddy Von Mueller. And last but not least, a special thanks to my editors, Peter-John Leone and Kathleen Zylan, who kept me on track when I tried to deviate and shared their valuable experience and objectivity when I was eager or impatient.

1

Track facts: what, where, and when

A limping theropod dinosaur, based on footprint evidence from the Jurassic of North Africa. Artwork by Edward Von Mueller.

There is no branch of detective science so important and so much neglected as the art of tracing footsteps.
– Sir Arthur Conan Doyle, *A Study in Scarlet* (1891)

Farmers ploughing their fields have dug up many intriguing but insignificant rocks. However, when Pliny Moody unearthed a fossil footprint in 1802, he did more than discover an unusual piece of New England bedrock. Though no one knew it at the time, he had found the first substantial evidence of dinosaurs.[1]

Sixty years later, after Professor Edward Hitchcock had devoted an entire career and almost a thousand pages of literature to documenting a large collection of these fossil tracks, they were still not recognized for what they were. The phenomenon that we know as dinosaurs had been defined solely on the basis of bones. Dinosaur skeletal remains commanded rapidly escalating scientific attention and popularity throughout the latter part of the nineteenth century.

A precedent was set. Bones were beautiful, highly prized and the basis for scientific study. Tracks, on the other hand, were troublesome, tricky, and too rare to be important; they were merely incidental by-products of dinosaur existence. This value judgment has prevailed until very recently and still persists in the minds of those uninitiated to tracking. But attitudes are changing. The dinosaur revival frequently focuses on evidence for dynamic dinosaur activity, and tracks provide information about dinosaur speed, social behavior, and ecology. Scientists have awakened to the fact that fossil footprints have been overlooked and ignored for too long.

If we just open our eyes, a new picture emerges. Far from being rare, tracks are ubiquitous. Strata from the age of dinosaurs are replete with billions upon billions of footprints. Tracks outnumber bones by several orders of magnitude and provide a vast new storehouse of valuable data to apply to our interpretations of dinosaurs. This book explains the whys and wherefores of some of these data and follows the trail of dinosaurs out of Sir Arthur Conan Doyle's fantasy Lost World into the mainstream of late twentieth-century earth science.

Before the reader can fully appreciate the meaning of dinosaur footprint evidence, some basic facts about dinosaurs and about the origin and distribution of their tracks are needed. Most people know the main groups of dinosaurs either by their popular names and representatives, or by their more technical classification. In Figure 1.1 we present a simple family tree showing the main groups of carnivorous and herbiv-

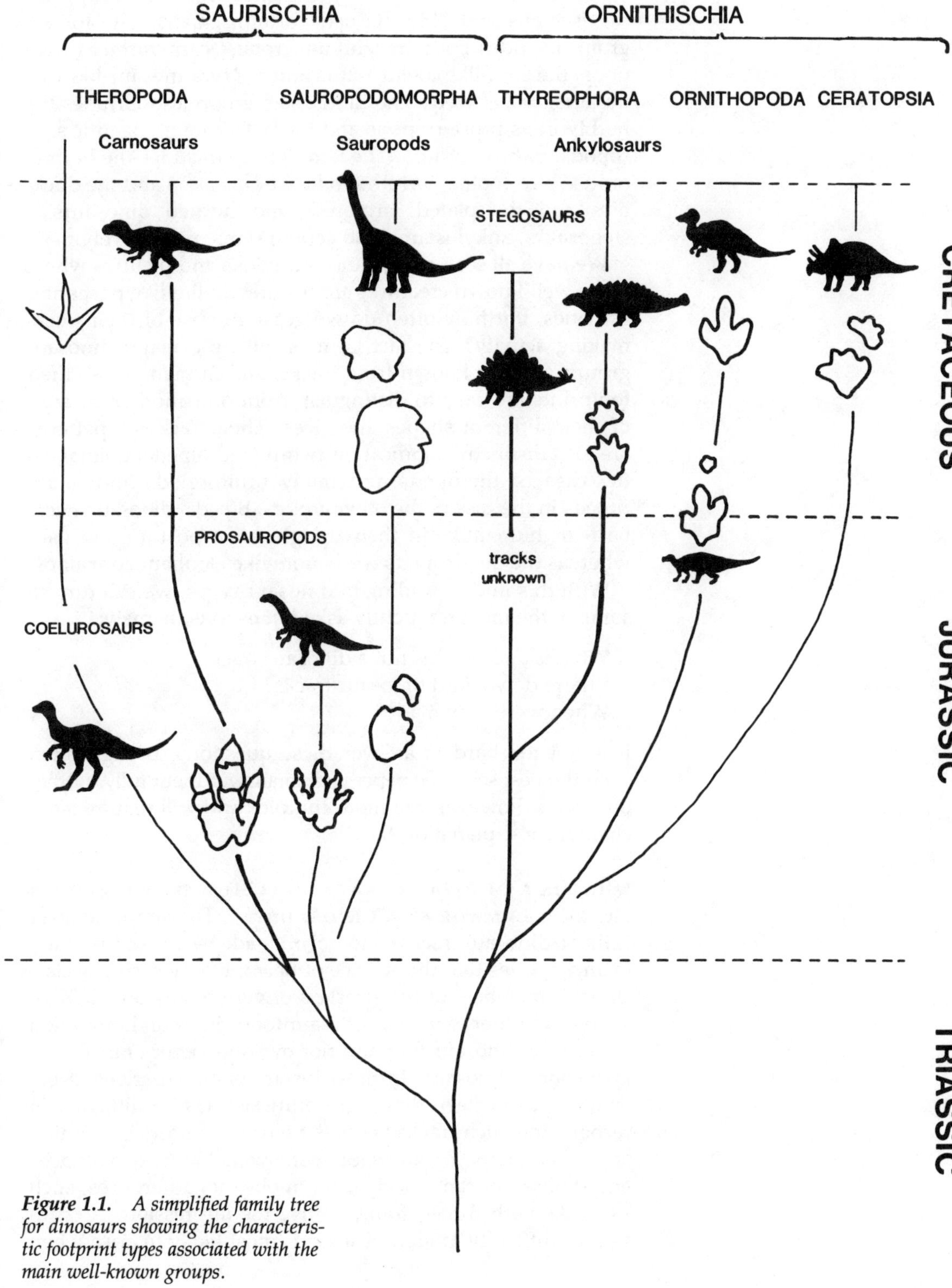

SAURISCHIA

ORNITHISCHIA

THEROPODA   SAUROPODOMORPHA   THYREOPHORA   ORNITHOPODA CERATOPSIA

Carnosaurs   Sauropods   Ankylosaurs

STEGOSAURS

PROSAUROPODS

tracks
unknown

COELUROSAURS

CRETACEOUS

JURASSIC

TRIASSIC

Figure 1.1. *A simplified family tree for dinosaurs showing the characteristic footprint types associated with the main well-known groups.*

orous dinosaurs traditionally classified as "lizard-hipped" saurischians and "bird-hipped" ornithischians. The former group includes both graceful and robust carnivorous theropods, the emu-like coelurosaurs and the tyrannosaur-like carnosaurs, respectively. The saurischian group also includes the herbivorous prosauropods and the well-known gigantic sauropods, or brontosaurs. The latter group includes the herbivorous "bird-footed" ornithopods like *Iguanodon* and the duckbills and the plated, armored, and horned dinosaurs – stegosaurs, ankylosaurs, and ceratopsians, respectively.[2]

We have all seen many dinosaur books and pictures where these well-known creatures are restored in life-like poses and activities, but how often do we get a glimpse of their track-making activity? The fact is, most of these major dinosaur groups are also known from tracks, and in many cases their footprints are easy to distinguish from one another because of their different shapes and sizes. Their trackway patterns are also distinctive, indicating two-footed bipedal animals in the case of theropods and many ornithopods and quadrupeds in the case of the other groups. Bipedal dinosaurs were bird- or humanlike in their upright erect posture and gait, whereas the quadrupeds were more like elephants or rhinos.

With this understanding of dinosaur types, we can turn to some of the most frequently asked questions, namely:

What is and what is not a dinosaur track?
Where do we find dinosaur tracks?
When were they made?

It is not too hard to answer these questions, though often even the best scientific papers fail to answer them fully. (Other questions, however, are more complex and will require a full chapter for explanation.)

WHAT IS A DINOSAUR TRACK? OR "HOW DO WE KNOW A TRACK WAS MADE BY A DINOSAUR?" The simple answer is that a dinosaur track is a footprint made by any of the many extinct species that the science of paleontology correctly classifies as members of the subclass of vertebrates called Dinosauria.[3] In other words, a dinosaur footprint is an impression made by a dinosaur foot and not by some other animal mistaken for a dinosaur. In most instances an experienced paleontologist or keen dinosaur enthusiast has no difficulty in recognizing such tracks because he or she knows something about the shape of dinosaur feet. Even the most inexperienced observer can quickly discern obvious differences, such as tracks with three-, four-, or five-toe impressions. Confusion or difficulty in identification is most likely to arise when particular dinosaur tracks resemble those of nondinosaurian trackmakers, such as birds, mammals, or nondinosaurian reptiles like crocodiles, pterosaurs, or turtles. As we shall show in later chapters, this confusion is not uncommon. Problems also arise when the tracks are very indistinct or poorly preserved.

Track researchers should be forgiven for making mistakes. Whereas an experienced tracker of modern animals, say, an Australian aborigine or Kalahari bushman, is unlikely to misidentify a trackmaker because he repeatedly comes upon animals he is tracking and so learns to differentiate their prints, those who attempt to classify prints made by extinct animals lack such an indisputable reference. Moreover, since hunting and tracking have ceased to hold life-and-death importance for us, scientists may on occasion be guilty of elementary errors in footprint identification. Rather than leading to empty food plates at the end of a day's hunting these mistakes have spawned scientific controversy, at times stimulating but often pointless. This is the nature of all human intellectual endeavor and the tracking of dinosaurs has been no exception.

In this urban age probably few of us can identify the different mammal, bird, reptile, and amphibian tracks found in the areas in which we live. In fact, track identification is not a particularly difficult task, but it takes practice and experience. In the same way that we are illiterate if we can't read a written language. We are illiterate as trackers if we can't read what footprints have to tell us. This is no less true for a dinosaur tracker. Practice reduces error and leads to a more organized, experienced approach. "Telling someone how to track is analogous to telling someone how to play the piano."[4] One learns primarily from experience.

Once bitten by the tracking bug, we can take a serious interest in what footprints tell us. We can organize our knowledge systematically much as in guidebooks on modern species. This systematic approach has been adopted in Chapter 5, where distinct track and trackway types are illustrated and described separately, with specific examples. This nontechnical "field guide" should help direct the enthusiast to sites where representative dinosaur species made tracks.

Even though there are many dinosaur encyclopedias available,[5] no field guide is possible for the body fossil remains of dinosaurs. This is because complete skeletons are not found lying conveniently in rock exposures. Even experts find it hard to identify bones half hidden in the rock; skeletal remains must be excavated and removed from the field for preparation in museums. Only then can identification, reconstruction, and exhibition be undertaken. Tracks, on the other hand, may ap-

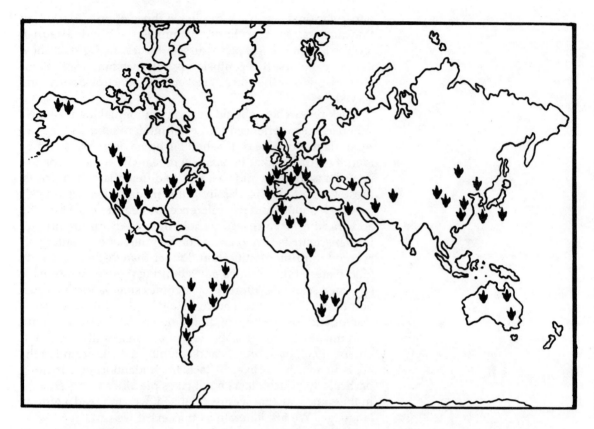

Figure 1.2. *Worldwide distribution of dinosaur footprint discoveries. About a thousand locations have been known to yield dinosaur tracks.*

pear in the field as clear and crisp as if the trackmaker had just walked by. With guide in hand and a little experience, one can readily identify many fossil dinosaur tracks.

WHERE DO WE FIND DINOSAUR TRACKS? Again, there are a number of straightforward answers to this question. We know tracks occur in sedimentary strata only, or at least only in strata that were sedimentary at the time of their original accumulation.[6] They occur, moreover, in sedimentary rock of terrestrial or continental origin; with the exception of tracks made in very shallow coastal waters, not exceeding dinosaur wading depth, no dinosaur tracks are known from marine deposits. They are in fact confined to terrestrial sedimentary deposits of Mesozoic age (Fig. 1.2). Accurate reports of dinosaur tracks in pre- or post-Mesozoic deposits are unlikely.[7]

As more has been learned about the distribution of dinosaur tracks, it has become clear that they are found preserved only in certain types of continental deposits. In most cases these are sediments that represented shorelines. These settings include beaches, tidal flats, and lagoon environments, along marine coasts, swamp and riverbank environments, lake

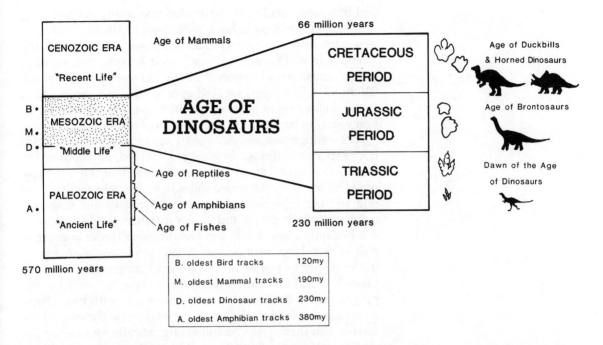

B. oldest Bird tracks	120my
M. oldest Mammal tracks	190my
D. oldest Dinosaur tracks	230my
A. oldest Amphibian tracks	380my

Figure 1.3. Geological time scale showing relative duration of the Age of Dinosaurs. Typical tracks from each time period are illustrated, along with the first appearance of tracks of other important vertebrate groups.

shorelines, and other settings where the soil or substrate underfoot was moist and receptive to track formation. In short, where we see tracks today. Some track-bearing deposits are confined to small areas, and some extend over very large regions. As we shall see, differences in the extent and distribution pattern of tracks provide geologists with valuable clues for reconstructing the configuration of ancient environments and charting the geography of the Age of Dinosaurs.

WHEN WERE THE TRACKS MADE? This question is most commonly asked in the hope of receiving an exact response, dating the trackmaking event precisely, say, at 100 million years. It is often not too hard for geologists to provide such a dating because a large proportion of Mesozoic rocks from the Age of Dinosaurs have been accurately dated by various geological methods.[8] In most areas we differentiate Mesozoic rocks into Triassic, Jurassic, and Cretaceous (see Fig. 1.3), and if the area has been well studied, it is possible to be considerably more accurate. Looking at a fine set of tracks, one may be asked, "When were they made?" or "How old are they?" One answers, "Between 99 and 100 million years old," and the enquirer is dumfounded: "You can tell that from the tracks?" The answer, frankly, is no; we already knew the age of the rocks from other dating methods.

It is worth noting that each geologic period is subdivided into epochs, for example, Lower, Middle, and Upper Triassic, and that these epochs are subdivided into clearly defined ages. Most epochs and ages have their own distinctive suites of dinosaurs and dinosaur tracks. When we consider that the whole Age of Dinosaurs ("age" used here in the popular, nontechnical sense) spanned about 170 million years (from 235 to 65 million ago), we find epochs, which average about thirty million years, and ages, which average about six million years, to be relatively short spans of geologic time.

As evolution progressed, age by age, epoch by epoch, different species of dinosaurs emerged, thrived, and became extinct. After careful study, it becomes clear that the track record is replete with distinctive suites of fossil tracks that succeed one another in meaningful ecological and evolutionary sequences (see Chapter 8 and 9). For instance, the Late Triassic exhibits only a few small, birdlike dinosaur tracks in association with a large number of varied nondinosaurian reptile tracks. Dinosaurs were at first a minor component of a reptile-rich world. By the Lower Jurassic epoch, however, the proportion of dinosaur tracks is very high, with more large forms accompanying the small varieties. The dynasty of dinosaur dominance was in full swing. Middle and Upper Jurassic strata generally lack small bipedal forms but yield instead an increasing number of sauropod or brontosaur tracks. This was the golden age of brontosaurs. Cretaceous epochs are generally characterized by an abundance of large ornithopod tracks; these are the ages of iguanodontids and duckbills.

The appearance of characteristic assemblages of tracks in different epochs allows us to trace the evolution of dinosaur communities through the footprint record, as we shall do in Chapter 9. Even in remote areas where the geology is poorly known, the presence of tracks can provide clues to the epoch in which the footprints were made. On several occasions geologists have used dinosaur tracks to help establish the correct age of important strata.[9] For example, if there were doubt about whether rocks were Triassic or Jurassic, the presence of tracks would almost certainly resolve the problem.

The footprint record suggests that the Age of Dinosaurs is well named. As the study of Mesozoic tracks has recently matured, it is increasingly clear that the vast majority of known tracks are unequivocally attributable to dinosaurs. The total number attributable to amphibians, lizards, crocodiles, turtles, pterosaurs, birds, mammals, and reptile-like mammals is relatively small. As we shall see, the distribution of these tracks, like those of dinosaurs, is strongly controlled by a number of factors. These include what we can call biological or paleobiological constraints, pertaining to behavior and evolution, and factors relating to the configuration of ancient environments. For example, various mammal-like footprints and other reptile tracks are relatively common early in the Age of Dinosaurs when the proportion of dinosaur tracks is relatively low. Later, toward the end of the Age of Dinosaurs, in the Cretaceous period, such mammal-like tracks are unknown, confirming the generally accepted paleontological belief that many of these animals had become extinct.[10] By way of contrast, true bird tracks are unknown or rare early in the Age of Dinosaurs. They first begin to appear abundantly in the Cretaceous (Chapters 8 and 9), shedding light on the evolution and diversification of our feathered friends at this time. Although we are now experienced at recognizing Mesozoic bird tracks, and distinguishing them from dinosaur footprints, confusion between bird and dinosaur tracks was a major problem and obstacle to the progress of the science of dinosaur tracking for many years. This confusion has only recently been resolved (Chapter 14).

It is not just evolution that determines where and when certain track types show up in the fossil record. We must also consider animal habits, in particular their preference for frequenting wet shorelines where they can make abundant tracks. Some tracks may be rare simply because certain animals did not hang out at the beach. Rock-loving and tree-perching birds provide an example; wetland-loving waders and waterfowl would be more commonly represented. We must also consider geological circumstances, which determine whether the tracks will be preserved after they have been made. If Mother Earth fails to cooperate, tracks will be destroyed with no regard for their potential usefulness to future generations of trackers.

We have begun our discussion of the types and locations of dinosaur tracks. However, before we look at tracks and sites in detail, we must turn to two fundamental questions. First, what do tracks mean in dynamic and biological terms? Second, how did they survive to get preserved as part of the fossil record?

2

The meaning of tracks

Two ostrich-like dinosaurs run along the Dinosaur Freeway, Cretaceous of Colorado. Artwork by Donna Sloan.

D.L. SLOAN '90.

Think of reading tracks and stories as a detective game.
— James Halfpenny, *Mammal Tracking* (1986)

We need to begin by asking what type of information is provided by individual tracks and trackways. This basic information will determine what interpretations can and cannot reasonably be made. Common sense tells us that tracks provide direct clues to size and locomotion, but not to features like skin color or brain capacity. Individual tracks or footprints (same meaning) are components of trackways (sometimes called trails) that show how an animal progressed for several consecutive steps. Trackways provide a more complete picture of animal locomotion than do isolated tracks, and they lend themselves more fully to interpretations of posture, gait, and speed. A single track or footprint is analogous to a snapshot, whereas a trackway is like a movie or a series of frames.

Trackways usually indicate immediately whether an animal is large or small, and whether it walks on two feet (bipedal) or four (quadrupedal). No one would confuse the tracks of an elephant with those of a horse or mistake a human footprint for a chicken track. Tracks also tell us about the shape of feet and the numbers of toes, details that are essential in identification of the trackmaker. We can also determine if the trackmaker walked erect or with a sprawling gait by looking at the width of the trackway. With experience, we can also easily distinguish running, walking, or hopping gaits. If an experienced modern tracker can tell a horse from a horse with a rider, a man carrying a suitcase from his unencumbered companion, surely we can distinguish different dinosaur groups.

We know that, in general, certain groups of dinosaurs, such as the carnivorous theropods and most ornithopods, were bipedal, whereas brontosaurs and plated, armored, and horned dinosaurs were quadrupedal. The former were like large birds; the latter, like elephants or hippos. To understand quadruped trackways we have to distinguish between the hind foot, or *pes*, and the front foot, or *manus* (sometimes called the forefoot or even the hand). The front feet were much smaller than the hind feet in most dinosaurs, and the difference is clearly seen in many trackways. However, the front foot–hind foot distinction may be obscure in fossil trackways, for several reasons. For example, some dinosaurs partially or completely overlapped their front foot impressions with their hind footprints, partly obscuring or completely destroying the front footprint only seconds after it was made. In such cases a quadrupedal animal could appear bipedal based on its trackway. Other dinosaurs carried very little weight on their front feet, so their forefoot tracks were both small and very lightly impressed. Like modern trackers, dinosaur trackers must be prepared to deal with trackways that are imperfect in certain details of their preservation. Imperfect tracks tend to be the rule rather than the exception. This is part of the challenge of the detective game.

Trackways of bipeds

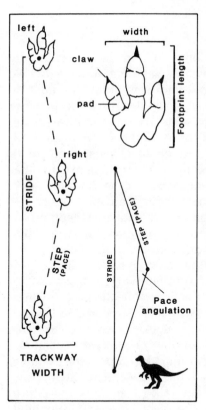

Figure 2.1. Trackway of a bipedal dinosaur showing step (pace), stride, and other measurable parameters that give clues about locomotion. Also shown are the main features of an individual track, including claw and pad impressions.

Trackways of habitually bipedal dinosaurs provide the simplest case for description and analysis as only the hind feet are involved. These animals walked rather like birds and human beings, never using their front limbs. The total width of the trackway from the outside of the left footprint to the outside of the right is called "trackway width" (Figure 2.1). This measurement helps define the erectness of the animal during locomotion: Bipedal trackways are always narrow because the trackmakers walked erect, putting one foot in front of another. If we imagine a line midway down the tackway, we note that bipeds stepped along, rather than straddled, this line, almost with the accuracy of a tightrope walker.

The angle between three consecutive tracks, right-left-right or left-right-left, is known as pace angulation. In bipedal trackways it is high, approaching 180°, or a straight line. Pace angulation is likely to vary depending on the length of the step, which is in turn a reflection of speed. We use the term *step* for consecutive footfalls (left-to-right or right-to-left footprints) and the term *stride* for consecutive footprints made by the same foot, right to right or left to left. If a dinosaur puts its feet exactly one in front of the other (pace angulation of about 180°), stride length is exactly double step length. If the feet are just slightly to the side of the midline, stride length is slightly less than two step lengths.

Other features of bipedal trackways include the degree to which the toes diverge outward or converge inward. This has been called positive and negative rotation, respectively, or a "duck-footed" or "pigeon-toed" condition in common parlance. Ornithopods are often slightly pigeon-toed, whereas carnivorous dinosaur trackways usually exhibit toes that point straight forward or diverge only slightly. Again, however, such inward and outward rotation depends to some extent on the speed of the animal at the time the tracks were made.

It goes without saying that any analysis of trackways should include an accurate description of individual footprints. In bipedal dinosaur trackways the foot is usually three-toed, or tridactyl (meaning three digits), with digits II, III, and IV being the ones represented. (The toes are counted as we would count our own toes, I through V, beginning inside, with our big toe, and counting outward from the midline.) Sometimes a four-toed (tetradactyl) condition is observed; occasionally a two-toed (didactyl) condition is reported, similar to the modern ostrich. One-toed (monodactyl) feet are virtually unheard of, and five-toed (pentadactyl) feet are uncharacteristic of bipeds, though quite common in quadrupedal dinosaurs. Trackers should beware of the pitfalls of misinterpreting tracks that reveal an unusual pattern of digit impressions. Too often we fall into the trap of supposing that a missing toe impression signals a new species. Some trackers have reported tracks with extra toes numbering six or even seven. These are not rare examples of multitoed (polydactyl) animals, however; they are simply examples of double prints! Occasionally there may be an unusual explanation, such as a maimed animal with a missing toe,[1] but usually an incomplete track is simply the result of poor preservation.

Trackways of quadrupeds

Trackways of four-footed dinosaurs can be systematically analyzed in much the same way as those of bipeds. We can record track size, overall trackway width, hindfoot-to-hindfoot and forefoot-to-forefoot step length, corresponding stride length, and pace angulation. We can also measure degree of footprint rotation. All stride and pace angulation values should be consistent throughout a given segment of trackway for the animal's locomotion to make sense. In other words, the measurements should reflect the dinosaur's ability to coordinate the movements of its hind and front feet. If they do not, it does not mean the dinosaur was inept; probably, our measurements require re-evaluation. We are learning locomotion from them, not vice versa.

In general we find that the trackways of habitual quadrupeds are broader than those of bipeds (Figs. 2.2 and 2.3). The hind and front footprints may straddle the trackway midline slightly. With four feet involved in support and locomotion, a narrow tightrope walker's trackway would be unstable and difficult to accomplish. Trackway width varies considerably in modern quadrupedal animals; hippos, for example, have wide trackways, whereas those of elephants are narrow. The differences are in part due to leg length, and in part to the speeds at which the animals move. Observations of modern animals also indicate that trackway width and pace angulation vary with speed. In other words, an animal generally plants its feet more widely on either side of the midline when standing still or moving very slowly, and it steps closer to the midline when moving forward. Compare, for example, the trackway of a walking hippopotamus with that of a run-

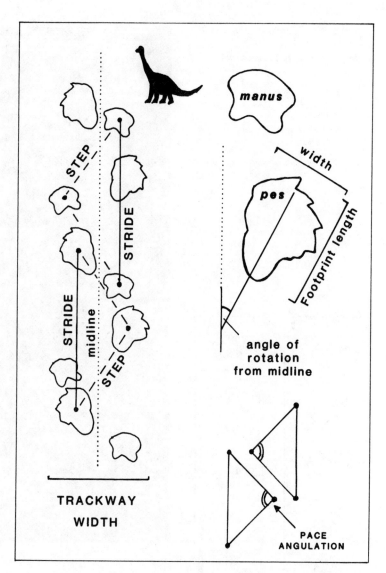

Figure 2.2. *Trackway of a quadrupe-dal dinosaur showing standard features that help trackers interpret locomotion.*

ning individual (Fig. 2.4). We only need reflect for a moment to realize that this difference allows for greater stability when standing still and greater efficiency in forward movement when in motion. The same simple principles apply to dinosaur locomotion and controlled the trackway width patterns preserved in the geological record.

Four-footed dinosaurs like the brontosaurs have marked differences between the shape of hind and front feet. Such differences in part reflect patterns seen in the anatomy of all dinosaurs (which almost always had smaller front feet), but they also reflect differences in limb length and in the distribution of weight on front and back feet. Sauropods probably

Figure 2.3. *Example of a bipedal dinosaur trackway. Note that left and right footprints are clearly visible.*

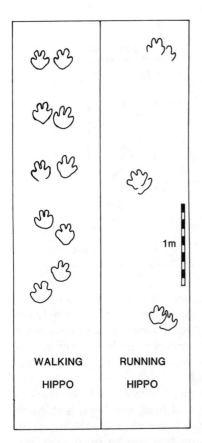

WALKING HIPPO   RUNNING HIPPO

1 m

Figure 2.4. *Trackways of a walking and running hippopotamus from northern Tanzania.*

carried most of their weight on their larger hind feet and only a minority on their small fore feet.[2] Their front legs were also shorter in many cases and hung straight down from their shoulders, in contrast to the back legs, which converged slightly beneath the body. Such differences help account for the fact that front footprints may be situated further out from the trackway midline than the hind footprints.

When dealing with quadruped foot shape, the front and hind feet must be treated separately, for they are usually very different. For example, brontosaur pes tracks are five toed in well-preserved footprints, whereas the manus tracks are hoof-like, showing no distinct digit impressions. By contrast, armored ankylosaurs and horned ceratopsians exhibit much less difference in the size of front and back feet. Both appear four toed, with little marked difference in shape (see Fig. 2.5).

Switching from two to four feet: optional and obligatory bipeds

Any animal that is not obliged to walk on either two or four feet at all times can be termed an optional or facultative biped. In other words it can switch from two-footed to four-footed progression and back. Few living species do this regularly, though lizards often use only their hind feet when running flat out. Kangaroos also put their front feet (and tails) down when progressing forward at a slow pace. Among dinosaurs, optional bipeds favored their hind feet and carried only a minority of their weight on their front feet. Large ornithopods like the iguanodontids and hadrosaurs provide the best examples of this facultative biped condition; their trackways indicate that they used both gaits. Again, trackers must be careful to check that such changes are real and not the result of imperfect preservation or changing patterns of pes/manus separation and overlap associated with changes in speed.

Walking flat footed or on tiptoes: plantigrade or digitigrade

Animals that walk flat footed, that is with their wrist and or ankle bones flat on the ground, are described as plantigrade, whereas those that walk on their toes, with wrist (metacarpal) and ankle (metatarsal) bones elevated, are described as digitigrade. The intermediate conditions can be referred to as semiplantigrade or as semidigitigrade.

In general, three-toed bipedal dinosaurs were digitigrade, toe walkers. This condition is typical of what are referred to as cursorial dinosaurs, or dinosaurs that are adapted for running. This adaptation is essentially similar to that seen in modern birds, which are thought by many to be descendants of the coelurosaurian dinosaurs. Digitigrade adaptations increase the leg length significantly because the metatarsal (ankle) bones are added into the overall leg length, raising the animals' center of gravity. A good analogy is seen in human athletes pushing off on their toes when running or jumping. Digitigrade dinosaurs were constantly on their toes. Occasionally such bipeds squatted down, leaving long metatarsal impressions on the substrate. In cases where the impression is clear, the length of the metatarsus can be estimated or measured. Such a posture could be called a plantigrade resting position. In some cases such traces are accompanied by small front foot impressions, supporting the interpretation of the animal as squatting or lowering itself toward the substrate in a crouching position (see Chapters 5 and 6).

A few trackways exhibit three-toed biped tracks with elongate metatarsal impressions. These have been interpreted by

Figure 2.5. Dinosaur tracks of different shapes and sizes. Note the variation in number of toes on hind and front feet. Tracks can be compared with trackway illustrations, Figures 5.1 to 5.4.

a small minority as human tracks because of their elongate shape and the heel-like appearance of the metatarsal impression.[3] Because such tracks date from the Age of Dinosaurs, they clearly cannot be regarded as of human or primate affinity, and serious study has shown them to be those of carnivorous bipeds moving with a plantigrade gait. Such unusual locomotion may represent some type of stalking activity, in which the body is lowered into a crouchlike posture. Alternatively, it may represent some form of display or threat behavior or simply be a result of walking over a soft substrate.

Unlike most bipedal dinosaurs, the quadrupedal varieties are generally large, rather ponderous heavyweights. They are referred to as graviportal, which means their skeletons are primarily adapted for weight bearing rather than for running

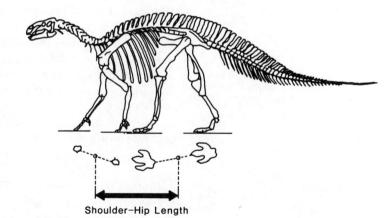

Figure 2.6. *Trackways of quadrupedal animals can be used to estimate hip-to-shoulder dimensions and hence give a good approximation of overall size.*

Shoulder–Hip Length

or agility. However, this does not mean they were flat-footed plodders. In general their limbs were erect and pillar-like, and their feet digitigrade. In fact, all dinosaurs had relatively upright postures and were derived from ancestors that had become erect toe walkers very early in the Mesozoic. Consequently many large quadrupeds inherited digitigrade, toe-walking, adaptations. In the case of brontosaurs the front feet are fully digitigrade, although the back feet tend to be slightly plantigrade. Trackways of stegosaurs, ankylosaurs, nodosaurs, and ceratopsians are not well represented, but what we have appears to indicate digitigrade to semidigitigrade adaptations.

Estimating the size of dinosaurs

Obviously, large tracks indicate large animals, whereas small footprints represent small trackmakers. As discussed in Chapter 3, small tracks may indicate small species, but they may also indicate babies or small individuals of a larger species. Trackers have devised various ways of using footprints to estimate dinosaur size. The simplest case involves bipeds where footprint length is about one-quarter to one-fifth of leg length or hip height (see Chapter 6). Thus if the track of a three-toed dinosaur measures 30 centimeters in length, we can estimate that it stood between 1.2 and 1.5 meters at the hip. Turning then to an appropriate known skeleton with legs of this length, we can complete our size estimate by measuring the head-to-tail length. In this example the animal would be at least 3 meters long.

We can do this with quadrupeds and estimate their hip height using hind footprint length. Moreover, with four-footed animals we have an alternative method at our disposal. As shown in Figure 2.6, it is possible to estimate the approximate position of both the hips and the shoulders of an animal on all fours. This is done by locating the two points midway between the hind footprints and the front footprints of the animal in quadrupedal pose. When we measure the distance between these two points, we derive an estimate of trunk length, or the distance between the hip socket (acetabulum) and the shoulder socket (glenoid). Once we have estimated this hip-to-shoulder, or gleno-acetabular, dimension, we can, again, go to an appropriate skeleton and easily estimate the animal's corresponding head-to-tail length.

Conclusion

Although most serious track researchers are cautious about reading too much into a single footprint, a footprint does provide information that can be useful in studying dinosaur foot shape. Many dinosaurs are known from incomplete skeletons that lack feet, and even when foot bones are preserved, the flesh and tissue anatomy, of course, is not. Only tracks can provide direct information on the size and shape of the fleshy foot pads. Nevertheless, however much information an individual track provides, it should always be viewed as a component of the overall trackway pattern, which gives the complete picture of the trackmaker's locomotion.[4]

Tracks and trackways tell us much about the posture, gait, and locomotion of dinosaurs. Features such as number of toes or pad configurations on hind and front feet can also be deduced from tracks more easily than from bones in most cases. For studies of locomotion, individual footprints are of secondary importance to the larger picture of the trackway configuration.[5] As we shall show in subsequent chapters, it is trackways that ultimately provide the fundamental insights into dinosaur locomotion and behavior.

3

Understanding track preservation

Reptile-like mammals (tritilodonts), small dinosaurs, and lizards congregate around a desert oasis. Based on trackway evidence from the Lower Triassic Navajo Formation of Utah. Artwork by Edward Von Mueller.

A thousand pyramids had mouldered down
Since on this rock thy footprints were impressed;
Yet here it stands, though since then
Earth's crust has been upheaved and fractured oft.
— Edward Hitchcock, *"The Sandstone Bird,"*
March 1, 1836

How have tracks been preserved hundreds of millions of years since they were made? This question is among those most frequently asked by laypersons and most frequently avoided by specialists describing dinosaur tracks. A simple answer, that the track-bearing layer was buried through geologic time, turning to rock without the destruction of the footprints, is inadequate because it does not explain how the tracks survived the burial and preservation process.

On first seeing fossil tracks, newly initiated trackers often imagine their own tracks on a sandy beach, susceptible to almost instantaneous destruction by the first wave to wash over them. This natural observation is highly pertinent because a large proportion of footprints are indeed destroyed soon after they are made. Each animal that lives is capable of making thousands, even millions of tracks in different environments, where the preservation potential varies considerably. Only a small proportion were preserved under favorable conditions.

Traditionally paleontologists have explained the footprint preservation process as follows. Tracks are made on a moist surface exposed to the desiccating, or drying, effects of the sun and atmosphere. As the wet sediment dries, it hardens, becoming more resistant to the erosive effects of the next tide or flood that brings in sediment to deposit the next layer and bury and preserve the well-defined footprints. Mud cracks, or sun cracks, as they are sometimes called, found on a track-bearing substrate prove that desiccation by the atmosphere was a factor in preservation. Undoubtedly this "cover-up" track preservation process or mechanism operated in the geologic past as it does today, but does it account for all preserved tracks? Too often, dinosaur trackers have used the cover-up explanation as an expedient to cover up their own lack of understanding.

Research has found another important mechanism. As feet sink into a soft substrate, they leave impressions in layers below the surface; footprints are made on underlayers. Such tracks are called undertracks, underprints, or ghost prints. Because they are already buried at the time they are made, their preservation potential is very high. The influx of subsequent tides or floods poses little threat of erosion to an impression already nestled a few centimeters below the exposed surface.

Serious consideration of undertracks constitutes three-dimensional rather than two-dimensional thinking. If we look at tracks on a layer of strata and imagine an ancient mudflat or beach, a flat surface, we are essentially confining ourselves to a two-dimensional realm of thought, considering only length and breadth dimensions. We must not forget the third dimension, depth. As layers of sediment accumulate, their thickness or depth introduces a very important third geological dimension: time. If a dinosaur steps on one layer and sinks through to leave its tracks on an underlayer five or ten centimeters below the surface, it may be leaving tracks on a layer that is hundreds or thousands of years older than the exposed surface. A thousand-year error on a million-year date may be insignificant in geological time (only one-tenth of 1 percent), but it is clearly significant in terms of interpreting dinosaur activity on a given day.

It is important to determine the layer on which the animal walked, for that is the stratum in which tracks are usually best preserved. As explained below, whether it is the track or the undertrack layer that is revealed depends on the inherent resistance of strata to present-day erosion and weathering. We cannot always be content with first appearances because they may be misleading.

There is more to understanding footprints than the outlining and measuring of a well-preserved trackway or the counting of toes and the cataloging of digit configurations. Tracks or trackway segments may present shapes that are less distinct than optimal textbook examples. The trick is to be discriminating, to glean information from evidence, whatever the quality.

In addition to enabling us to make an accurate description of foot shape, locomotion, posture, and gait, well-preserved tracks tell us that the substrate was neither too hard or too soft. Hard substrates are not at all conducive to preserving tracks, and soft, sloppy substrates produce tracks that are indistinct and hard to identify. Tracks made by the same animal can vary considerably in quality, providing an insight into the properties of ancient substrates. The evidence is not necessarily easy to find. Some tracks, representing once-firm substrates, are very shallow and recognizable only with optimum illumination. (Trackers often make their best observations of shallow tracks early in the morning or late in the day, when the angle of the sun's rays is best.) Other tracks are very deep, with indistinct outlines caused by the collapse of semifluid, easily deformed sediment.

We do not have to depend only on pristine trackway configurations. Even a poorly preserved brontosaur trackway may show features that are clearly brontosaurian. The same is true, at least in principle, for undertracks. If we know what we are looking at, we may be able to derive valuable census information without actually observing true tracks in the strict sense. This is the essence of good detective work; the conscientious paleontologist leaves no stone unturned.

Paleontologists recognize several different types of preservation. Fossil tracks may be petrified replicas of the original footprint impression, essentially molds of the feet. Strictly speaking, these are dynamic impressions, or molds of the foot in all phases of motion during the footstep. Some trackers have gone so far as to differentiate these dynamic components, as the touch down, weight-bearing, and push-off or kick-off phases of the footstep.[1] These tracks may be noticeably different from the impressions taken from the feet of lifeless museum specimens.[2]

If footprint molds are filled in by sediment, natural casts are produced (Fig. 3.1). These are essentially replicas of the feet formed by the overlying sedimentary layers. They are sometimes referred to as tracks in negative relief, as distinct from the positive relief of molds. Dramatic examples are sometimes found in coal mines, where footprint casts bulge down from the mine roof after the coal seam has been excavated. (Such large downward bulges and projections are hazardous to miners. There is at least one report of a miner being killed by charging headlong into a footprint cast.) Whether molds or casts are preserved is largely dependent on the resistance of the rock in the track-bearing and the infilling layers. If a soft mud fills in a track in a firm sand that later becomes hard sandstone the track will probably be preserved as a mold after the soft shale has been eroded out. Conversely, if sand fills a shale mold, it will be preserved as a natural cast. Where two tough layers are preserved, we may find both the molds and the matching casts.

As mentioned, undertracks, or ghost tracks, are formed by the impact of feet on underlayers below the surface on which the dinosaurs walked. A single footprint may produce several underprints stacked on top of each other. Tracks may also fill in as layers of overlying sediment accumulate, leading to an expression of the true footprints in the overlaying strata. Because they were made after the tracks were formed, they are not the direct, dynamic result of animal activity; they only indirectly reflect the track shape. Track infillings are generally obscure, so they are often overlooked and have rarely been studied.

NATURAL CAST

TRACK INFILLING

TRUE
TRACK

UNDERTRACKS

Figure 3.1. Top. *Footprint impressions, casts, underprints, and infillings, as they appear in rock strata.* Bottom. *Natural impressions (left) and natural cast (right) of a small Late Triassic three-toed dinosaur from Oklahoma.*

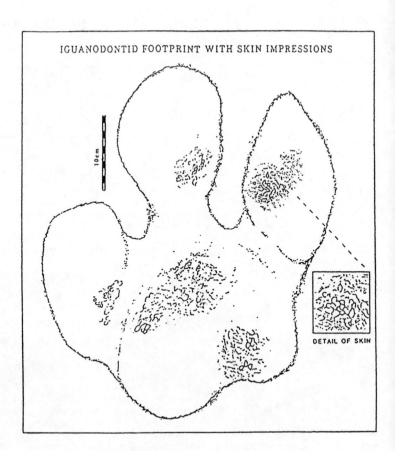

IGUANODONTID FOOTPRINT WITH SKIN IMPRESSIONS

10cm

DETAIL OF SKIN

Figure 3.2. A true footprint, preserving skin impressions.

Good preservation: true tracks

Any layperson with normal observational skills can recognize a well-preserved track. However, defining good preservation scientifically is not so easy. One of the most obvious criteria is the presence of skin or scale impressions, which confirm, without a doubt, that the track in question is a true track, but these impressions are surprisingly rare in dinosaur tracks. There are only a few examples from the entire latter part of the Mesozoic Era (Jurassic to Cretaceous).[3] (See Fig. 3.2.) Examples from earlier in the Mesozoic are mainly attributable to nondinosaurian reptiles.[4]

There are a number of reasons why skin impressions are rare in footprints. First, a footprint is made while the animal is in motion. The slightest movement or slippage can obscure or smear the fine texture of skin impressed in the substrate. This helps explain why skin impressions are commonly associated with mummified dinosaurs, which were rapidly buried after death. Their carcasses were stationary, allowing sediments to mold the finest detail without disturbance.

Other factors are also important in contributing to good

preservation. Skin impressions will only occur in sediment that has the right texture. Usually sticky mud, the consistency of potter's clay, is the best molding medium. However, after the mud has lithified to mudstone and is exposed at the surface, it is more susceptible to erosion and weathering than any other rock type. Thin muddy layers in which skin impressions are preserved are likely to be the first layers lost to erosion even if well-preserved footprints remain.

Track depth is also an important factor in the quality of preservation. Again, simple observation tells us that well-preserved tracks are usually moderately deep. A track that is extremely shallow is almost imperceptible because the substrate has not yielded to the force of the foot. At the other extreme, a very deep track indicates an excessively soft, or what geologists call an incompetent, substrate that provided little or no resistance. Often such substrates are very wet or loose, as in the case of dry sand, and they collapse into the footprint as soon as the foot is withdrawn. The moderate-depth footprint represents a balance between the strength of the substrate and the force exerted by the foot. Such a balance obviously depends on the condition of the substrate and the weight of the dinosaur. Well-preserved tracks of small chicken- and turkey-sized bipedal dinosaurs are typically only a few millimeters deep and rarely in excess of a centimeter in depth, whereas well-preserved brontosaur tracks range from 5 or 10 up to 20 or 25 centimeters in depth.

Other types of track preservation: undertracks

Some substrates are more conducive to good track preservation than others. Cohesive muddy sediment provides a good molding medium, especially if it is firm and not too wet or sloppy. In such a medium the outlines of tracks are clear and crisp, with an abrupt change in slope where the edge of the track intersects with the substrate. Crisp track outlines are very helpful when it comes to measuring track size. However, the ideal medium is not always encountered. Substrates vary in consistency, not just laterally as one moves from one area to another, but vertically from layer to layer or, more correctly, from layer to underlayer. An animal may step through 10 centimeters of water, an impossible medium for preserving tracks, or through 10 centimeters of soupy fluid mud, also an impossible medium, only to leave a perfect trackway in a cohesive underlayer. In such a situation the modern tracker could not follow the trail, but it would be clearly marked in the sedimentary geologic record. In theory it could be followed by draining off the water and sludge down to the firm substrate. The geologist is fortunate because time has removed the water and sludge and, if the circumstances were

FOSSIL VOLUME

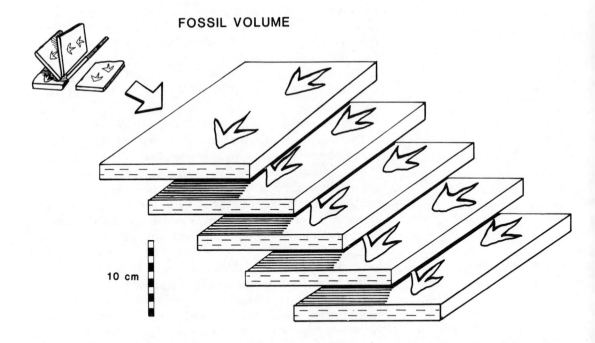

Figure 3.3. *The principle of under-print formation, as illustrated by Ed-ward Hitchcock's* Ichnology of New England *(1858). Note the underprints on at least four underlayers below the true tracks.*

right, revealed tracks that were invisible at the time they were made.

Archaeological excavations in Japan have revealed abundant well-preserved human footprints in 1,500-year-old rice paddies. These footprints would have been quite invisible at the time they were made because they were buried beneath water and sludgy mud.

Sedimentary layers often consist of alternating layers of mud and sand, mud and silt, mud and limy mud, or other combinations. In studying undertracks, it is important to note the thickness of the layers relative to the depth of the footprints and also the appearance of the same footprints in successive layers. Obviously, thinly layered sediment will produce several super-imposed undertracks. Edward Hitchcock vividly demonstrated this point in 1858 by illustrating a stacked series of tracks in a thin sequence of layers bound together like the pages of a book; he called this his fossil volume.[5] As you turned back the pages, you could see how the footprints were transmitted as underprints in successively deeper layers (see Fig. 3.3). this is analogous to writing on a sheet of paper and making carbon copies underneath – or, better, to leaving colorless impressions on several sheets in a pad. In such a series the footprint's margin becomes increasingly obscure in successive layers. In other words, the change in slope between the substrate and the footprint impression is usually most abrupt on the surface where the animal walked. Shallow tracks with diffuse margins are typical of underprints.

Because modern trackers do not peel back sediment layers to see what undertracks look like, these geological insights hold little significance for them. However, geologists must examine these layers if they hope to interpret tracks correctly. The simple fact is that these are often the only layers we see in the rock record. As suggested later, some of the most serious dinosaur footprint misinterpretations have risen from failure to distinguish undertracks from the true tracks.

Two examples will illustrate the types of problems encountered. The first involves the quadrupedal dinosaurs, which carried more weight on their hind feet. If their hind feet sank more deeply, than their front feet into the substrate, the underprints of their hind feet were also transmitted to deeper strata. As mentioned, it is the weight borne relative to the size of the foot that determines how deeply a foot sinks. Feet with a small surface area can sink deeply (think of stiletto heels). A quadruped can appear bipedal because its front footprints did not sink as deeply as its hind. (Alternatively, the front feet may sink in deeper, as in the case of some brontosaurs, whose front feet are smaller and spread the weight less widely. Any animal can appear to be walking on several layers representing different times.) The small dinosaurian trackmaker *Agialopus* provides an illustration of this phenomenon (Fig. 3.4). Because it carried little weight on its front feet, its tracks are typical of small carnivorous dinosaurs from the early Mesozoic, which were almost always bipedal. We find its front footprints only lightly impressed at higher levels, and never at the deepest levels reached by its hind feet. As discussed later dinosaur front feet may have sunk deeper than their back feet in some cases.[6]

The second example involves tracksites where two or more different types of dinosaur have left tracks or undertracks on the same surface. In practice as well as in theory the larger, heavier trackmakers leave deeper tracks, so only their footprints penetrate as underprints in the underlayers. The bigger they are, the deeper they sink. Two or three species may leave true tracks on what was once the surface, but only the larger species leaves its mark on the underlayers (Fig. 3.5). As we shall see, the proportion of small tracks from the Age of Dinosaurs is rather low, suggesting another bias affecting our discoveries: Big tracks are simply easier to find.[7] (A similar problem occurs in the study of dinosaur skeletal remains. The larger bones of larger individuals are less rapidly destroyed before being buried by sediment. Moreover, these bones are easier to spot. So a double bias may favor the pres-

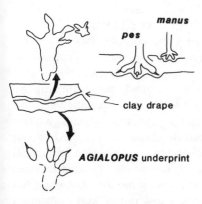

Figure 3.4. The front and hind foot of a small dinosaur (maker of Agialopus tracks) sank to different depths in the same substrate. Thus the same animal left different track patterns at two different levels in the strata.

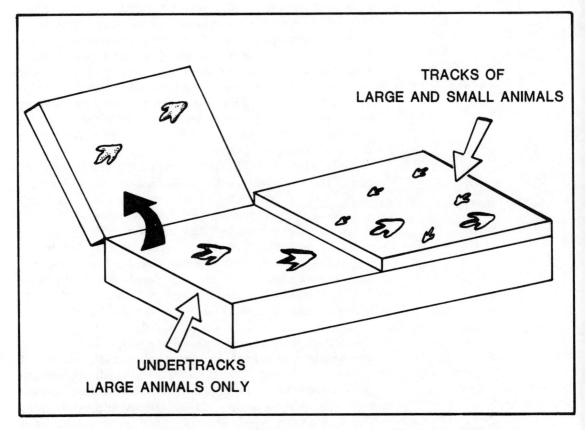

TRACKS OF
LARGE AND SMALL ANIMALS

UNDERTRACKS
LARGE ANIMALS ONLY

Figure 3.5. *Underlayers may leave a biased sample of the dinosaur community by favoring the preservation of the larger tracks.*

ervation and discovery of large bones over small ones. Paleontologists try to avoid this bias by deliberately looking for small bones. This is often done by sieving sediments. The result is sometimes a bias in favor of small fossils.)

We would be oversimplifying the complexities of preservation if we did not take two further points into consideration. First, undertracks do not necessarily reflect heavy animals; they may reflect relatively soft substrates at the time the trackmakers passed by. When we find considerable variation in the depth of the tracks made by the same species, we can conclude either that the substrate consistency varied within the tracksite area or that it changed over time as individual trackmakers passed by. Second, footprints of different sizes may not represent different species. Some dinosaurs, particularly the three-toed carnivores, show remarkably consistent foot shape over a considerable size range (Fig. 3.6). Small tracks may represent juveniles of a species.[8]

At least in some cases the track record may be biased against the preservation of baby tracks as well as against those of small species. Again we run into a parallel paleontological problem. For a long time people believed that baby dinosaurs

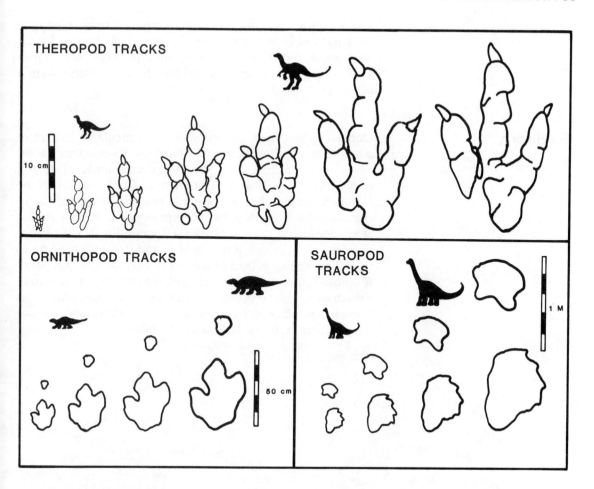

THEROPOD TRACKS

10 cm

ORNITHOPOD TRACKS

SAUROPOD TRACKS

1 M

50 cm

Figure 3.6. Size range in various types of dinosaur tracks often indicates the presence of juveniles and subadults with fully grown adults. In the past, theropod tracks of different sizes were usually interpreted as the footprints of species of different size.

were rare. However, recent work by Jack Horner and others in nest-site regions has shown that this is not true.[9] We just did not know where to look or how to account for preservational biases. The debate over the rarity of the skeletal remains of baby dinosaurs prompted at least one dinosaur tracker to suggest that tracks in fact indicate a distinct absence of juveniles.[10] Although there are relatively few juveniles' tracks in many regions, this observation does not take into account the preservational biases discussed by Horner and others in recent years; it is a good example of a conclusion based on negative evidence. Small tracks are probably rare because they are not preserved, not because baby dinosaurs were rare. After all, all large dinosaurs were once small!

We now know of several tracksites that reveal the presence of both medium-size and large individuals inferred to belong to the same species. As discussed in Chapter 5, there is debate over whether small theropod tracks represent small species or babies. However, we have always assumed the small and medium-size brontosaur and ornithopod tracks belong to the same species as the larger tracks with which they are found. Very small tracks of these dinosaurs are very rare or unknown, suggesting, again, a preservation bias rather than a lack of juveniles.

Conclusion

Understanding track preservation is particularly important when studying fossil footprints. Although the modern tracker may not be concerned with the three-dimensional, geological perspective, the paleontologist must be. Consideration of the factors that control preservation is essential if we are to understand the biases against the preservation of true tracks, skin impressions, and small tracks. Though the problems may be complex, they are fundamental, and it is encouraging to note that the biases being revealed by current research parallel those known in other fields of paleontological research. This means that precedents exist for recognizing these as known biases that need not throw off our interpretations unduly. We can proceed with some confidence.

4

Discovery and documentation

A herd of five partially grown sauropods walks along a Late Jurassic shoreline, based on trackway evidence from southeastern Colorado. Artwork by Doug Henderson.

It happened one day about noon. . . . I stood like one thun-
derstruck, or as if I had seen an apparition . . . for there was
exactly the print of a foot – toes, heel and every part of a foot.
                              – Daniel Defoe, *Robinson Crusoe* (1719)

There is always the chance of discovering dinosaur tracks by
sheer luck. After all, they are often large and conspicuous
and would be hard to miss on the right rocky outcrop. Of
course, such an outcrop would have to represent sedimentary rock deposited in a continental environment during the
Age of Dinosaurs. Such rock can be found almost anywhere
– in mountains, river valleys, mines, quarries, in road cuts,
or along sea cliffs.

Many fossil footprints discoveries have been made in rural
or remote areas by local people who lack any formal paleontological training. The first dinosaur tracks ever found were
discovered this way, in 1802, when Pliny Moody of South
Hadley, Massachusetts, ploughed them up on the family farm.[1]
He and his contemporaries believed them to be bird tracks.
Other tracks are discovered as the by-product of rock-quarrying
operations, rather than by pure chance. An example would
be the abundant dinosaur tracks discovered in coal mines
throughout Colorado and Utah.[2] Similarly, other Lower Jurassic track sites in New England have been discovered during excavations, notably the Dinosaur State Park site at Rocky
Hill, Connecticut, which was discovered during highway
construction operations.[3] A survey of about 150 tracksites in
the Rocky Mountain region revealed that about two-thirds were
found in natural exposures, the rest in quarries, mines, and
roadways (Fig. 4.1).

Once discovered, a site should be quickly reported to the
landowner, local paleontologists or other experts, and authorities so that they can determine how the site is to be studied or managed. In theory the local experts and authorities
can proceed with the business of documenting the site for
science and posterity. This is the ideal type of discovery and
documentation scenario in which the community members
bring forth valuable information about a site in a cooperative
and responsible fashion. Fortunately this type of scenario is
quite common.

Sometimes, however, a smooth transition from discovery
to documentation does not occur. One of the first obstacles
may be the remoteness of a site. If the local discoverers cannot find experts nearby, it may be difficult and impracticable
to bring in anyone to investigate. The site will then remain

34

WHERE TRACKS ARE FOUND

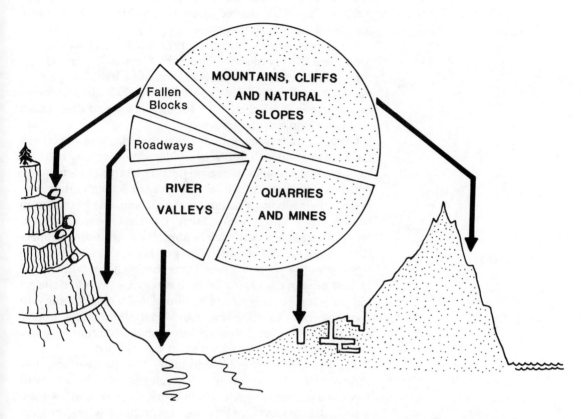

Figure 4.1. Dinosaur tracks are found in many different natural and artificial terrains.

"known" but languish in obscurity indefinitely. A good example is the Late Jurassic Purgatoire Valley site in southeastern Colorado. It was discovered and brought to the attention of the local community in 1935, briefly reported in the local press and through popular science outlets, and then forgotten about for the next fifty years.[4] Several other sites have suffered similar fates. for example a remote, large, Early Jurassic site in Arizona was discovered and photographed in 1934 by Roland T. Bird (see Chapter 13). After this, it was not studied, and though "discovered," its precise location was unknown. Again, it was fifty years before the site was rediscovered.[5]

There are a few other spectacular examples of remote frontier tracksites. Near Chacarilla Oasis in Chile hundreds of Late Jurassic dinosaur tracks are exposed on steeply inclined surfaces in the rugged terrain of the Andes. According to the geologists who discovered them, they are only accessible to trained mountain climbers.[6] Another inaccessible dinosaur tracksite has been reported from a snowbound area high in the Swiss Alps. In the rugged canyons of the Colorado and Green rivers on the Colorado Plateau, giant house-sized boulders fall from precipitous cliffs of Mesozoic rock. Dinosaur tracks are often discovered on exposed surfaces where layers have tumbled and split apart.

We cannot blame a site's remote location in all cases. There has been a history of scientific apathy toward dinosaur tracks throughout much of the past century and a half. The perception that tracks are of little value has clearly had much to do in many cases with the failure to convert exciting discoveries into worthwhile documentation.

Other problems sometimes beset the adequate documentation of sites after their discovery. Landowners may be wary for personal or political reasons about letting students, scientists, or experts on their land. Rightly or wrongly, they may believe their dinosaur resources are too valuable to release to someone else's jurisdiction. Such situations are unfortunate, and although they may make entertaining after-dinner anecdotes, they are not conducive to good science. One example from recent years involves a Texas landowner who believed he had human tracks alongside those of dinosaurs at a site on his property. For this reason he hoped to develop the site into a commercial attraction and would not allow access to any scientists who might be outspoken against his claims.[7] Clearly such an approach is unscientific, to say the least.

Petty politics may enter into many a battle over who has jurisdiction at any given site. The scientist may have to wait weeks or months to get permits or permission to investigate a particular discovery. However, it would be wrong to view landowners and land management agencies as necessarily obstructive or villainous in such situations. They have primary responsibility for the site and may wish to take time to arrange access according to protocol or some schedule other than that preferred by the scientific investigators. It is worth adding this cautionary note. The scientist does not have a God-given right to immediately investigate every interesting discovery brought to his or her attention. The paleontological community reverberates with amusing, sometimes hair-raising anecdotes of encounters among scientists, landowners, and bureaucrats. Discoverers and documenters need to exercise caution when stepping on someone else's turf. Common sense and courtesy can be rewarding, whereas brashness and self-importance can result in a close encounter with the business end of a gun barrel.

In other instances, where sites are accessible, they may be subject to imminent destruction by commercial quarrying operations. Scientists then have to embark on a rescue mission to document the site and, of course, do so in cooperation with the commercial operations. A large gypsum quarry in Arkansas, known as the Briar site, provides a good example. Geologist Jeffrey Pittman and many of the quarry workers had been puzzled by flat layers of strata pock-marked with large potholes. Equipment operators and truck drivers tried to keep these large indentations covered with mud to avoid excessive wear and tear on vehicles constantly bouncing in and out of these tiresome holes. When Pittman realized they were tracks he had to move quickly to make maps and replicas before quarry operators excavated and destroyed the whole layer. His rescue mission was a success, resulting in substantial documentation and the acquisition of rubber and fiberglass molds of many of the best trackways.[8]

It is often desirable to monitor quarry operations. In some cases tracks are uncovered and covered again as rock is removed and open cast excavations subsequently filled in with rubble or other tailings. The process is analagous to natural erosion that uncovers and covers or destroys outcrops. Patience and long-term monitoring help to reveal the maximum amount of available information.

Assuming that a newly discovered site is accessible for study and documentation, the scientific method imposes certain demands. Documentation procedures should be orderly and systematic, resulting in information or data suitable for a comprehensive report or scientific paper for publication. At the very least, the investigators should record or attempt to record the following[9]:

Location of the site
Geological formation and age of track-bearing layers
Names and descriptions of different tracks and trackways
Orientation or direction of tracks and trackways
Measurement of track size, depth, step, stride, etc.
Relationship of tracks to sedimentary rocks
Sampling procedure (tracks or replicas collected for museums)

This essential information will form the basis for further discussion and interpretation of the site. In order to protect important discoveries, it may be necessary to withhold publication of the exact location of certain sites. However, all the other data can usually be recorded and published without reservation.

Most dinosaur trackers begin the documentation process by making a map of the site. This requires cleaning the site of all debris so that the tracks can be clearly seen (Fig. 4.2). It may also require repeated visits to the site to observe the track-bearing layer under different conditions of illumination. Optimum illumination is important for making the best obser-

Figure 4.2. *A typical track-bearing layer after it has been swept clean, ready for mapping and replication. Early Jurassic site, eastern Utah.*

vations and obtaining good photographs. Experience shows that tracks show up best when they are illuminated by low-angle light. Such light usually strikes horizontal surfaces early or late in the day. On inclined surfaces the best light may be at odd hours.

A map is a convenient way to show the exact location and orientation of every track on a given layer. As shown in the Jurassic and Cretaceous dinosaur tracksite examples in Figures 4.3 and 4.4, symbols can be used to illustrate various different types of track made by herbivores or carnivores, bipeds, or quadrupeds.[10] Maps are user friendly for the researcher and report reader and easier to interpret than a data sheet full of coordinates and measurements. For example, a

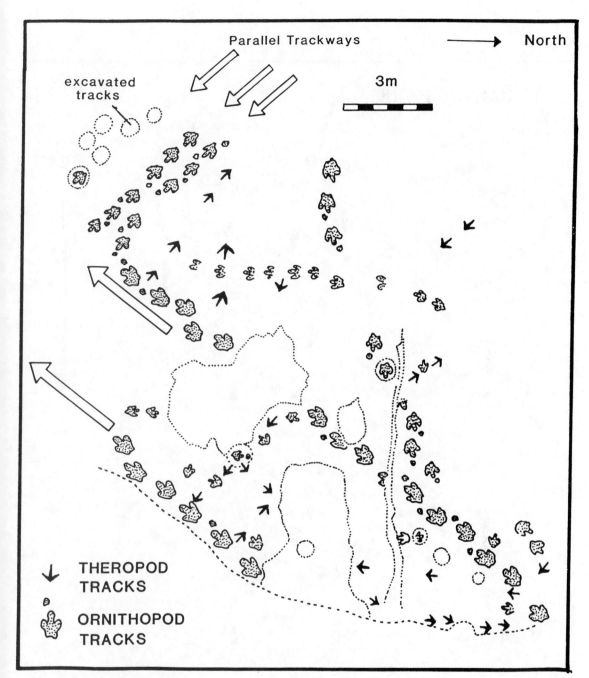

Parallel Trackways

North

excavated tracks

3m

THEROPOD TRACKS

ORNITHOPOD TRACKS

Figure 4.3. *A typical map of dinosaur tracks, showing Cretaceous trackway patterns from Dinosaur Ridge, near Denver, Colorado. Note the different symbols used for herbivorous dinosaur (ornithopod) and carnivorous dinosaur (theropod) tracks. Map also shows scale, orientation, and damaged (excavated) areas of the outcrop.*

39

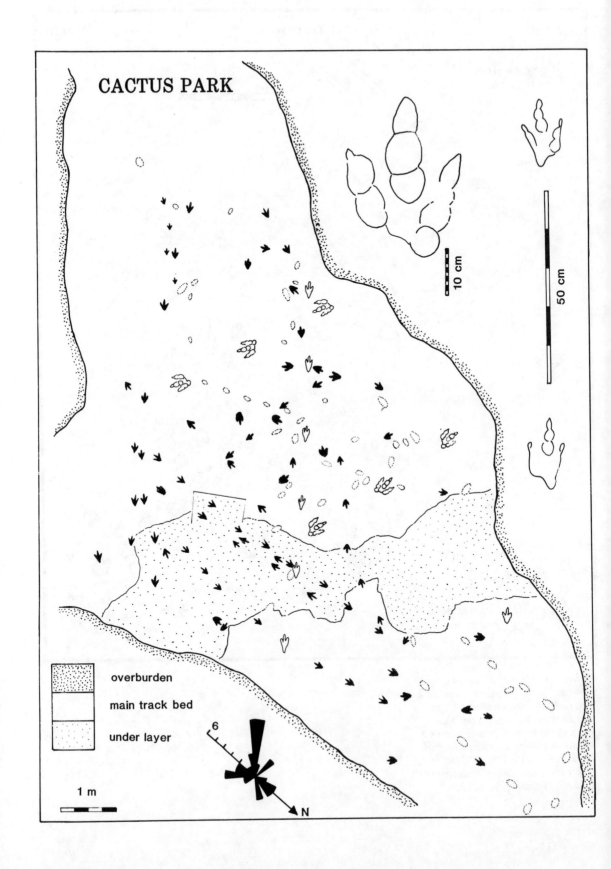

CACTUS PARK

10 cm

50 cm

overburden

main track bed

under layer

1 m

6

N

map makes it immediately possible to see the direction in which various trackmakers were traveling and to sense whether they were moving individually or as part of a larger group.

Systematic measurement of the size and depth of footprints and the length of steps and strides can begin after the map is made. With map in hand, we can number each trackway as it is measured and documented in detail. The measurements needed have been given in Chapter 2; however, it is worth pointing out that reliable data can be obtained without measuring every footprint, step, and stride. The reason is obvious: Within segments of a trackway, repeated footprints and strides are likely to have the same or very similar dimensions, unless the trackmaker is slowing down, speeding up, or crossing from one type of substrate to another, so the measurement of a few consecutive steps should provide sufficient reliable data on an individual dinosaur. The investigator can use his or her own judgment in deciding on the number of measurements needed to produce reliable average footprint and step dimensions. Where trackways are long and well-preserved, measurement of four or five consecutive tracks and intervening step–stride lengths should suffice. Where trackways are shorter and poorly preserved, it may be necessary to measure everything in order to get reliable data and minimize error.

It is important to document details of track preservation, particularly in the case of imperfect preservation. Are the tracks true tracks or undertracks, casts or molds? Are they deep or shallow? Do toe, claw, or pad impressions show up clearly? What is the type of sediment in which the tracks occur, and what type of sediment fills them in? Are there associated fossils or other geological features worth noting at the site or in the immediate vicinity?

Once these questions have been addressed adequately, the paperwork phase of the documentation process is almost complete. A written record has been established for posterity. However, one other phase of documentation may be desirable and even essential. This involves making permanent casts or replicas and in some cases collecting actual specimens for museum collections and exhibits.

Unlike bones, which cannot be properly seen without excavation, tracks are often best left in place. Removing individual footprints usually destroys the integrity of trackways. However, if the site is in danger of being destroyed by erosion, vandalism, or further excavation, excavating tracks and trackways and placing them in a permanent safe repository, such as a museum, may be desirable. This is a worthy rescue or salvage mission, and it also makes the specimens available for study by future generations of researchers.

Figure 4.4. *Map of Lower Jurassic tracksite at Cactus Park, western Colorado. Note the large track type (Eubrontes) and small type (Grallator), both shown in detail top right. Symbol is also used for indistinct tracks, and trackway orientations are shown in diagram (lower center).*

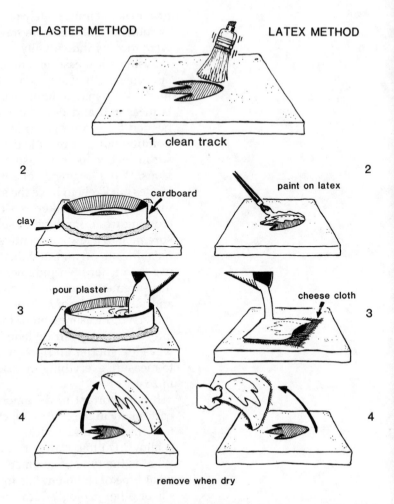

PLASTER METHOD   LATEX METHOD

1   clean track

2

cardboard

clay

pour plaster

3

paint on latex

2

cheese cloth

3

4

4

remove when dry

Figure 4.5. *How to make dinosaur track casts from plaster of paris (left) and latex rubber (right).*

There are not many examples of large-scale excavations of dinosaur trackways. Probably the best example is that of the excavation of a large segment of brontosaur trackway by the American Museum of Natural History from a Cretaceous site in Texas. The excavation, supervised by Roland T. Bird in 1940 was on a massive scale, and it took years to complete the task of transporting the material to the museum and reassembling it.[11] Most other excavations have been on a more modest scale involving the removal of smaller tracks and trackway segments, often from layers of strata much thinner than those encountered by Bird.

An alternative method of preserving footprints, is to make casts or replicas for museum collections without removing or damaging the original tracks and trackways. Casts are easily made by pouring plaster of paris into the original footprint impression, leaving a "negative" cast after it has hardened and been removed (Fig. 4.5). This technique of cast making is quite simple. First, you build a retaining wall of clay or some other suitable material around the tracks, so that the fluid plaster mix does not flow away when poured. The track must also be cleaned and, depending on the rock type, coated with liquid soap or some other effective "separator" so that the plaster does not stick to or bond with the rock as it solidifies. If a separator is not used, the plaster may adhere to the rock so well that it will be impossible to remove. The advantage of the plaster of paris method is that it is quick and the materials (water and plaster) are cheap. The disadvantages are that plaster is brittle, rather bulky and heavy, especially for large track casts, and cannot be used effectively on a track-bearing surface that is steeply inclined.

Common alternative methods of casting and replicating include the use of latex and silicon rubber. Latex rubber is normally applied to tracks in several thin layers and reinforced with cheese cloth. Once the latex cast has been dried and cured to a tough, flexible skin, it can be peeled off and used to make a replica. This can be done either immediately or later in the lab at the researcher's convenience. The disadvantage is that latex is not rigid. It is also somewhat expensive and only dries quickly under warm, dry conditions. Silicon rubber has most of the advantages of latex but is prohibitively expensive if used in large amounts.

Sometimes original specimens are so valuable that decisions are made to collect and preserve them with little regard for their size or the labor involved. We recently excavated a three-ton slab because it revealed both bird and dinosaur footprints on the same surface (see Chapter 9). Because such slabs are so large, they must be broken – often along natural fractures lines – and then stuck back together again in the museum (Fig. 4.6).

Once specimens have been collected and casts and replicas have been made, the final phase of documentation involves giving the specimens catalogue numbers so that they can be permanently deposited in an appropriate museum or repository. Here they are accessible to scientists and enthusiasts who might not otherwise be able to gain access to the original site, especially if many years have elapsed since its discovery or if the tracks have deteriorated through erosion.

Of course, the final phase of documentation is not really complete until all the pertinent information is assembled in a scientific publication. Such articles present the basic data and also allow the author to present interpretations and formulate hypotheses. Papers that present interpretations without accompanying data are speculation, not science. As shown in Chapter 13, speculative interpretation of footprints are often easily falsified. Converting the geological evidence into an enduring scientific contribution is an important aspect of preservation. Although tracks are vulnerable to destruction

Figure 4.6. Track-bearing slabs are reassembled after excavation. Cretaceous specimen from Colorado (see Fig. 9.8).

by weathering, erosion, and sometimes vandalism, well-conceived scientific contributions cannot be so marred. In many instances they will endure and remain valuable longer than the tracks themselves.

Discovery and documentation of a site may appear a simple two-step process. However, science does not rely primarily on blind luck for its discoveries. The more we learn about tracksites, the more we are in a position to predict the type of deposits in which they occur and hence where to look.

Experience has taught us that tracks are often found on a flat, freshly exposed rock surface. Such surfaces are found extensively only in coastal cliffs, river valleys, deserts, high mountains, or excavations where vegetation and soil cover are sparse or nonexistent. A systematic search of such outcrops ensures a greater probability of success than does random prospecting. This is especially true when searching geological formations in which tracks are already known.

It has recently been found that certain large tracksites extend for considerable distances in thin layers or zones (discussed in Chapter 12) that are often associated with the boundary layers between geological formations. This makes the track-bearing layer easy to locate in areas where tracks have not previously been reported. Recently many new sites have been discovered simply by predicting where these extensive track-bearing zones will appear in a surface rock outcrop.[12] The researcher may earmark likely localities on a geological map before leaving for the field. When he or she arrives at the predicted tracksite the probability of success is likely to be very high. On one occasion I found four tracksites in a single afternoon using this predictive method.

So the two-step discovery and documention process is really a cycle or a positive feedback loop: Discovery leads to documentation, which in turn leads to predictions and further discovery. More important, it is effective science.

5

Classification: a field guide to dinosaur tracks

An Early Jurassic ornithopod sits out a rainshower, based on trackway evidence from New England. Artwork by Edward Von Mueller.

In the form of ten . . . "commandments," a standardized procedure for the description and illustration of fossil vertebrate footprints is proposed.
— William Sarjeant, *Dinosaur Tracks and Traces* (1989)

The branch of dinosaur paleontology dealing with tracks and their classification is known as dinosaur ichnology (from the Greek *ichnos*, meaning "trace"). Specialists or ichnologists in the field have classified various dinosaur tracks into morphotypes (particular shapes), morphospecies, ichnospecies, and ichnogenera. Clearly dinosaur footprints are not exempt from the influence of classification schemes, which are designed to make life orderly and simply, but paradoxically make it complex and verbose. This branch of systematic classifications is sometimes called just that, systematics.

In keeping with the well-established traditions of general biology, dinosaur remains are classified according to the well-known binominal system of zoological nomenclature introduced by Karl Linnaeus in the latter part of the eighteenth century. Thus a dinosaur can be accorded the same scientific status as a living animal and be classified as belonging to the genus *Tyrannosaurus*, it can further be given a species, or trivial, name *rex*; reading the names in tandem, we get *Tyrannosaurus rex*, or in abbreviated form *T. rex*.

The same principle has been applied to the naming of fossil footprints, though with some modifications. In theory and in practice it is impossible to determine exactly which species made a particular set of tracks. For this reason we cannot use the same species designations reserved for fossil bones and record "footprints of a *Tyrannosaurus rex*"; the skeptics would rightly demand proof. Usually the best we can do is refer to "footprints that are *Tyrannosaurus*-like." If such tracks are distinctive, the scientifically inclined dinosaur tracker may wish to erect a formal name for such tracks. In the case cited above, the name *Tyrannosauripus* has been proposed. The system works well because in this case the name (ichnogenus) means "tyrannosaur foot," or "tyrannosaur-like foot." The name implies an association with *Tyrannosaurus* but is different enough from the genus to avoid confusion. Even if the name implies a track made by *Tyrannosaurus*, it does not, strictly speaking, imply a particular species of trackmaker. Another example, ichnogenus *Brontopodus* obviously suggests a brontosaur track. Other names are less specific, for example, *Tetrapodosaurus* ("four-footed reptile"), but still convey something of the trackmaker's morphology. We note in these examples that footprint names frequently incorporate -*pus*, -*podus*, and so on, indicating "foot." However, some tell us nothing about the characteristics of the animal that made them. For example, *Caririchnium* indicates traces or tracks from the Carir Basain area of Brazil. Another example is *Brasilichnium*,[1] but this track genus is not attributable to a dinosaur. The ending -*ichnium* tells us we are dealing with a trace fossil but does not expressly denote a footprint.

In theory there should be one named ichnogenus and ichnospecies for each genus and species of dinosaur responsible for leaving distinctive tracks. This situation is more or less what we find in field guides to modern tracks. Each species leaves distinctive footprints. *Distinctive* is the operative word here: As explained in Chapter 3, many dinosaur species, especially those belonging to particular families, had similar, even almost identical feet, although other parts of their anatomy, on which their classification usually hinges, may have been significantly different. Consequently, we might expect body fossil species names to outnumber those of trace or footprint species, and this is exactly the situation we find in present-day dinosaurian paleontology. We know of almost nine hundred dinosaur species reported from the body fossil record but only a few dozen valid and widely used footprint species.[2] As with many branches of systematics, track classification is in need of considerable revision and updating.

The purpose of this chapter is not to catalogue all morphotypes or ichnospecies exhaustively, but rather to illustrate the major groups and their distinctive characteristics. As in Chapter 3, we begin with bipeds both for convenience and for systematic and evolutionary reasons.[3]

Theropod trackways (Fig. 5.1)

The Theropoda represent a very large order of dinosaurs comprising all the carnivores and a few toothless varieties. They are often conveniently, if somewhat artifically divided into the infraorders Carnosauria and Coelurosauria (see Chapter 1), broadly defined as larger, robust *Tyrannosaurus*-like forms and smaller, gracile *Coelophysis*-like forms, respectively.[4] The differences between the tracks of these two genera and their allies are quite obvious in most cases. Whereas a large tyrannosaur track might measure half a meter or more in length and sink ten centimeters into a relatively firm substrate, a small coelophysid track would resemble that of a turkey and penetrate little more than half a centimeter into a substrate of similar consistency. At the extreme small end of the spectrum, a coelurosaurian track might appear no larger than the footprint of a small pigeon or blackbird.

For paleontological and historical reasons the most abun-

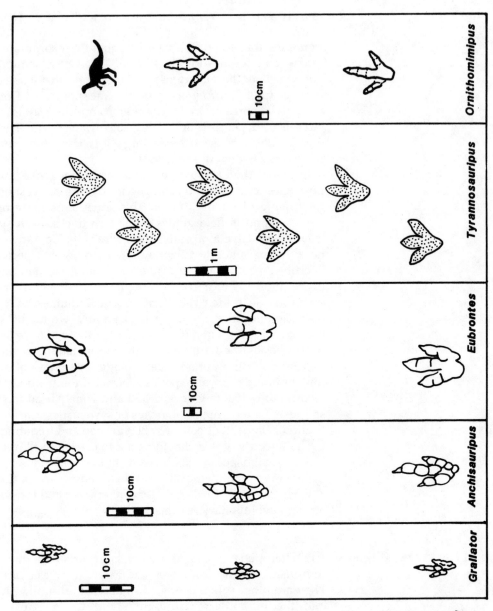

Figure 5.1. Theopod dinosaur tracks. Notice that the Tyrannosauripus tracks are drawn in cartoon style because the original trackway was never properly illustrated. The true trackway was much narrower, like the other trackways.

Grallator

Anchisauripus

Eubrontes

Tyrannosauripus

Ornithomimipus

10cm

10cm

10cm

1 m

10cm

dant and best-known theropod tracks originate from early Mesozoic deposits, particularly those of Lower Jurassic age.

Grallator (meaning "similar to the tracks of the Grallae, or heron–stork family")

Tracks of this type are very abundant, particularly in the Early Mesozoic. They were first documented by Edward Hitchcock in 1858[5] in his description of various turkey-like or bird-like tracks reported from the Connecticut Valley as early as 1802 and were initially called Ornithichnites, or "bird traces." Well-preserved tracks of the Grallator type exhibit characteristic pad impressions allowing every detail of the foot's jointing and anatomy to be clearly seen. Most people accept that these tracks were made by bipedal dinosaurs similar to coelurosaurs, the probable ancestors of birds.

Anchisauripus (meaning "Anchisaurus foot")

Another well known Grallator-like footprint, it is essentially three-toed but sometimes shows the trace of a fourth toe in the heel position. It was described by Richard Swann Lull in 1904[6] as the probable track of Anchisaurus, a small, mainly bipedal prosauropod dinosaur discovered in Massachusetts in 1857 and described by O. C. Marsh in 1885. According to Colbert, Anchisaurus was probably no more than two to two-and-a-half meters in length. Many experts now believe that Anchisauripus tracks are attributable to Coelophysis-like theropods, rather than to prosauropods. This shows the danger in referring a track to a particular genus of trackmaker.

Eubrontes (meaning "true thunder")

A contemporary of Grallator, the much larger ichnogenus Eubrontes, named by Hitchcock in 1845,[7] apparently represented one of the first of the larger carnivorous dinosaurs. The trackmaker was bipedal and probably a robust carnosaur, not a lightweight coelurosaurian type. Some interpret Eubrontes as large examples of Grallator, but most trackers regard it as a distinct ichnogenus.

Ornithomimipus (meaning "ornithomimid-like foot")

Described in 1926 by Charles Sternberg[8] on the basis of Late Cretaceous trackways in Western Canada, Ornithomimipus is a classic example of a trackway that has been linked to an inferred trackmaker, in this case the toothless coelurosaurian dinosaur Ornithomimus, relative to the better-known "ostrich dinosaur *Ornithomimus*, relative to the better-known "ostrich dinosaur," *Struthiomimus*. In many respects, particularly the slender toes and narrow trackway, *Ornithomimipus* resembles *Grallator*. It is one of only a few later cretaceous theropod trackways named to date.

Tyrannosauripus

This name was given in 1971 by the distinguished vertebrate ichnologist Hartmut Haubold,[9] to large three-toed tracks discovered in a Late Cretaceous coal mine in Utah in 1924. Despite the name, this track type is not well known and may not even have been made by a tyrannosaur. Like many coal mine tracks it could have been made by a duck-billed dinosaur. This uncertainty is largely because the original trackway, on which the description was based, is no longer accessible for study. As shown in Figure 5.1, the trackway was never depicted accurately.

Prosauropod trackways

Prosauropods, or sauropod predecessors, are a controversial group of dinosaurs. Anatomists argue over their relationships to brontosaurs and the accuracy of the name itself. Trackers debate what their footprints look like. Certain dinosaur trackers have proposed that prosauropods made bipedal, three-toed trackways like *Eubrontes*, whereas others have suggested they made four-toed tracks like *Otozoum* (meaning "giant animal"). Doubt was cast on the three-toed interpretation when Donald Baird described a quadrupedal four-toed trackway, *Navahopus* from the Jurassic Navajo Formation of Arizona, as a fine example of a trackway of a small prosauropod. He said that even a skeptic could clearly see the similarity between the tracks and the prosauropod foot skeleton.

Over the past century dinosaur trackers have debated the affinity of *Otozoum* tracks. Some say the tracks were made by prosauropods; others suggest some type of crocodile; yet others have avoided suggesting a trackmaker altogether.[10] In my opinion, the large size of the tracks and the two-footed nature of many of the trackways argue against a crocodilian. The prosauropod interpretation is the most reasonable, but as indicated in the notes, it is an example of a fossil footprint that has proved controversial.

Sauropod trackways (Fig. 5.2A–C)

Sauropod, or brontosaur, trackways are known from several dozen scattered localities worldwide, ranging in age from Early and Mid-Jurassic to Late Cretaceous. They are also known from Asia, Europe, Africa, and South America, but Texas is the most famous area for brontosaur footprint localities.

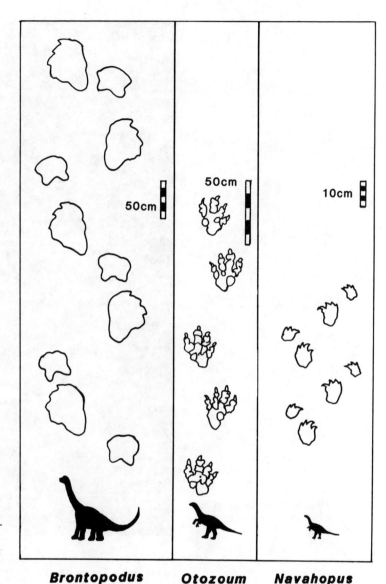

Figure 5.2A. Sauropod and prosauro-pod trackways. Otozoum was one of the first reptile trackways discovered in North America, yet the identity of its trackmaker still remains controversial.

Brontopodus   **Otozoum**   **Navahopus**

Brontopodus (meaning "thunder foot")

Brontopodus is an example of a trackway type that has been unequivocally assigned to a well-known trackmaking group. *Brontopodus*-like tracks were first discovered in the 1930s along the Purgatoire River in Colorado. They were brought to the attention of the scientific community by John Stuart Mac-Clary; a plucky invalid who could not visit the site himself, MacClary enlisted the help of friends to photograph the spec-tacular tracks. He reported the tracks, through the scientific grapevine, to Roland T. Bird, who examined them in 1938 on

Figure 5.2B. *Shallow brontosaur tracks on a cliff face near Rancho del Rio, Colorado.*

his way to Texas. Bird recognized that they could have been brontosaur tracks but found them less spectacular than examples he himself later unearthed in Texas. Bird devoted much time to the documentation and excavation of brontosaur tracks at several Texas sites, and in his unpublished writings proposed the name *Brontopodus*. This name was not published until 1989, when a new generation of Texas dinosaur trackers began reexamining the material. These workers, including Jim Farlow and Jeff Pittman, formally named the tracks *Brontopodus birdii*, a fitting tribute to one of science's greatest contributors.[11]

Brontosaur or *Brontopodus*-like tracks are known from localities throughout the world. The height of brontosaur ascen-

Figure 5.2C.   Deep brontosaur tracks near Moab, Utah (see also Fig. 6.4).

dancy was in the Late Jurassic, when *Brontosaurus* (strictly, *Apatosaurus*) itself walked the earth. The famous American Museum exhibit, with parallel brontosaur and carnosaur trackways, uses a Late Jurassic *Apatosaurus* to simulate the trackmaker, even though the tracks are Late Cretaceous in age. A fuller explanation is given in the section on R. T. Bird in Chapter 14.

Ornithopod trackways (Fig. 5.3)

Ornithopods, particularly the large iguanodontids and duck-billed hadrosaurs of the Cretaceous, were responsible for making a variety of trackways, both bipedal and quadrupedal. Their feet were generally three-toed, like theropods, but their tracks were much broader and their step shorter.

Anomoepus (meaning "unlike the hind and front feet")

First described by Edward Hitchcock in 1848,[12] *Anomoepus* is probably the best-known example of an early ornithischian (or ornithopod) trackmaker. It dates back to the Early Jurassic. It also provides a good example of a trackway type that varies considerably, not least because the trackmaker was a facultative biped, sometimes walking on two feet and at other times on four. Further details are given in the section on behavior (Chapter 6).

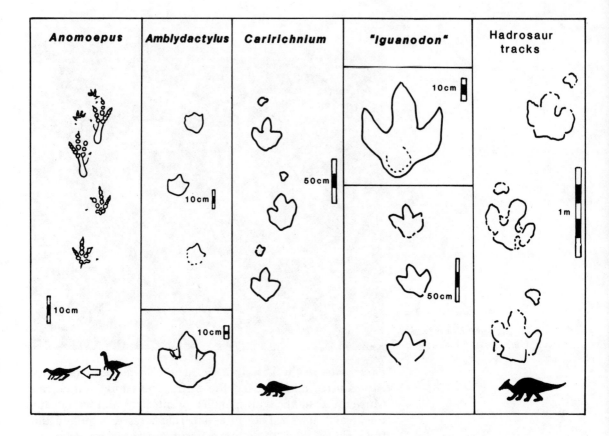

Anomoepus	Amblydactylus	Caririchnium	"Iguanodon"	Hadrosaur tracks

Figure 5.3. *Ornithopod trackways. Note the change from bipedal to quadrupedal in* Anomoepus, *the only Jurassic trackway. Note that all other trackways are Cretaceous and that there is a mixture of bipedal and quadrupedal progression.*

Amblydactylus (meaning "blunt toed")

This very broad three-toed *Iguanodon*-like trackway was first described by Sternberg in 1932,[13] based on Lower Cretaceous trackways from Western Canada. *Amblydactylus* trackways sometimes show hooflike footprints indicating that the trackmaker sometimes walked on all fours. In this respect, *Amblydactylus* resembles *Caririchnium*, a quadrupedal ornithopod.

Caririchnium (meaning "track from the Carir Basin of Brazil")

Caririchnium was named by Giuseppe Leonardi in 1984.[14] He originally interpreted the trackmaker as a stegosaur, but later, after seeing similar tracks in Colorado, where the front footprints were better preserved, he was convinced that the trackmaker was an ornithopod.

    Caririchnium clearly shows remarkable differences between large, broad hind feet and diminutive, hooflike front feet. At least one *Caririchnium* track, from Colorado, exhibits skin impressions (see Chapter 3). The trackmaker could have been an iguanodontid or a very early representative of the duck-billed Hadrosauridae family.

Iguanodon (meaning "iguana tooth")

Iguanodon is a scientifically inappropriate name for a footprint, although it clearly implies that the trackmaker was the Lower Cretaceous dinosaur *Iguanodon*. For reasons given above, most dinosaur track specialists agree that it is inadvisable to use an actual dinosaur name for a track. A term like *Iguanodonichnus*, proposed by Casamiquela and Fasola in 1968,[15] is more consistent with the principles of ichnology. However, because of the historical circumstances of finding these large three-toed ornithopod tracks in the same deposits as *Iguanodon* bones, along the south coast of England, they came to be known simply as *Iguanodon* tracks as early as the 1860s. In fact, a strong case can be made for regarding them as the first tracks correctly and unambiguously assigned to any dinosaur. The fact that they are also undoubtedly of *Iguanodon* or iguanodontid affinity underscores the accuracy of the original designations.

Most *Iguanodon* trackways indicate bipedal animals, though a few suggest four-footed progression and indicate that the small *Iguanodon* forefoot sometimes made contact with the ground. However, it appears the *Iguanodon* trackmaker may have been less inclined to walk on all fours than its two relatives, *Amblydactylus* and *Caririchnium*.

Hadrosaur tracks

Various duckbill dinosaur, or hadrosaur, tracks are abundant in Late Cretaceous deposits, but they have yet to be adequately and systematically described. A number of names have been proposed but as is sometimes the case, these names, including *Hadrosaurichnus*[16] (meaning "Hadrosaur track"), have been proposed for poorly preserved footprints, whereas well-preserved tracks that have more recently come to light remain unnamed. As trackmaker identification is one of the main objectives of tracking, it is sometimes desirable to avoid existing classifications if they deal only with indistinct material. In time the more distinctive, unequivocal hadrosaur tracks will be adequately described in the scientific literature.

Trackways of plated, armored, and horned dinosaurs (Fig. 5.4)

Despite an abundance of well-known quadrupedal ornithischian[17] skeletons attributable to stegosaurs, ankylosaurs, ceratopsians, and others, trackways of these groups are poorly known. Proven stegosaur tracks are unknown, and those of the other groups are rare. To date we know of only two prob-

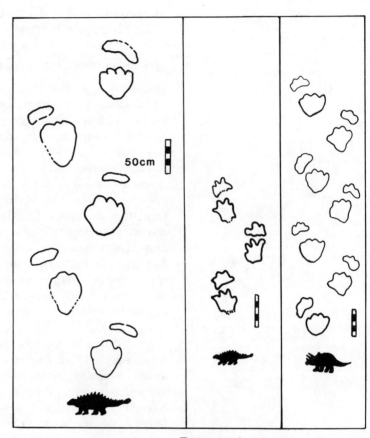

Figure 5.4. Trackways of quadrupedal ornithischians. Left: a probable ankylosaur from England; center: a Tetrapodosaurus from Canada; right: a ceratopsian from Colorado.

Tetrapodosaurus

able ankylosaur trackways and a few probable ceratopsian trackways. None of these rare yet important trackways have been described in great detail.

Tetrapodosauropus (meaning "four-footed reptile tracks")

Tetrapodosauropus is an example of a rare but well-preserved quadruped trackway, described by Charles Sternberg from the Lower Cretaceous of Canada in 1932.[18] According to Ken Carpenter, it probably represents the trackway of an armored dinosaur of the nodosaur family. Recently a much older – Late Jurassic – quadrupedal trackway was discovered in England. Paul Enson suggested that it could have also been made by a "nodosaurid ankylosaur" (or by a sauropod).[19]

Ceratopsian trackways

Although the ceratopsians, horned dinosaurs like *Triceratops* and its relatives, were very common during the Late Cretaceous, their trackways appear to be surprisingly rare. This may be because they usually frequented environments with dry substrates, where tracks were not easily made or preserved. Most of the possible ceratopsian trackways that have been reported are known only from a few isolated tracks or indistinct examples that lack appropriate names. The example in Figure 5.4 was discovered in Late Cretaceous deposits of Colorado[20] which are traditionally assigned to strata known as the *Triceratops* biozone. In other words, the tracks occur in the same strata as the skeletal remains of ceratopsians (see Chapter 8).

Conclusions

Most of the major groups of dinosaurs have distinctive trackways that fall into one of the dozen or so named types illustrated in this chapter. Unfortunately the more than five hundred additional names that clutter up the literature are largely redundant. Many were named to describe individual footprints whose relationship to an overall trackway pattern was, and still is, unknown. Perhaps this is a blessing in disguise because it means we can ignore or reject many of the more obscure names. Common trackway names like *Grallator* describe numerous trackway types that create a large catalogue of mainly unpronounceable names. Similarly, *Brontopodus* will probably prove adequate to describe a good proportion of known brontosaur trackways.

Assuming that we know what features to look for in various distinctive dinosaur trackways, the next questions to ask are (1) Where can we see good examples, and (2) after identifying the probable trackmaker, what else can we learn from the trackways? The first question is answered in Appendix A, "Where to Visit Dinosaur Tracksites." The second topic is the subject of the next chapter, on individual behavior.

6

Individual behavior

A Cretaceous theropod running. Based on trackway evidence from Texas. Artwork by Michel Henderson.

Fossil footprints are strangely evocative, because they directly capture the movements of once-living beings. They are the closest thing to dinosaur "motion pictures" that we have.
– Gregory Paul, *Predatory Dinosaurs of the World* (1988)

Now that we can recognize various different types of dinosaur trackways, it is time to turn our attention to what individual trackways can tell us about behavior. Behavior can be broadly defined and describes many different forms of activity. When an individual animal stands, sits, walks, accelerates, runs, slows, turns, hops, jumps, or even limps, it is engaging in different forms of activity or locomotion that constitute behavior. The reasons for such behavior may not always be clear. For example, an animal may run because it is being chased or, alternatively, because it is in pursuit of prey. In one sense the behaviors are similar – running. In another sense they are different – fleeing versus attacking. Similarly an animal may sit or crouch to rest, hide or stalk. It may limp because of a recent wound or because of a bone disease. Unfortunately, motives and reasons are rarely discernible from fossil evidence.

If we examine the track record, although thousands of dinosaur trackways are known, only a very small proportion, less than 1 percent, indicates unusual behavior. All the trackways illustrated in the previous chapters are those of walking animals. Moreover, at present we only know of a few trackways attributable to running dinosaurs, mainly theropods and small ornithopods.

This is not surprising when we consider that most animals spend virtually no time running, because of the high energy cost. In other words, we have yet to find trackway evidence that proves that large ornithopods, brontosaurs, horned, or armored dinosaurs ever ran. Such evidence agrees with various independent calculations of dinosaur size and weight that have classified the group into categories of running ability, ranging from good runners to ponderous heavyweights. This work, by Walter Coombs,[1] identifies small and medium-sized bipeds (theropods and ornithopods) as good runners (cursorial), larger bipeds (large carnosaurs and large ornithopods) as moderately good runners (subcursorial), and large quadrupedal ornithischians and sauropods as ponderous heavyweights (graviportal). This does not mean that large quadrupeds could not run, just that they were not designed to excel as runners. Not all dinosaur authorities fully agree with

Coombs in all details, but his classification has considerable merit as a general model.

The best example of a running dinosaur trackway is probably the widely spaced sequence of footprints reported by James Farlow[2] from Lower Cretaceous limy mudflats in Texas. The theropod responsible for making the trackway had a foot length of 37 centimeters and was covering 5.3 meters per stride (Fig. 6.1). Farlow estimated its speed at about 40 kilometers per hour, using a formula proposed by R. McNeil Alexander. This velocity is about the same as that achieved by an olympic sprinter covering 100 meters in 9.9 or 10 seconds. Of course, unlike the olympic runner moving flat out, the dinosaur responsible for the Texas trackway may have been traveling at less than its maximum velocity. Few other authentic examples of trackways of running individuals are known. The best one I know of is a theropod that left its tracks in the mudflats of an Early Cretaceous lake in Korea. All other known examples of running dinosaurs are dealt with in the next chapter.

In the late 1970s there was considerable debate over the speeds attained by dinosaurs. Much of this debate took place on the pages of the prestigious British journal *Nature* and revolved around dubious reports of a "mystery" duckbill dinosaur trackway with exceptionally long strides (see Chapter 13). R. McNeil Alexander proposed a formula that loosely translated into a simple rule of thumb.[3] A dinosaur trackway indicates walking if the step is less than four times foot length, and running if greater than four times foot length. Using this formula the vast majority of dinosaur trackways indicate slow progression in the range of about 1 to 3 kilometers per hour. As suggested earlier, this is what we might expect.

Subsequent work by Tony Thulborn suggested some important modifications to Alexander's popular approach.[4] Thulborn rightly drew attention to the need to modify speed estimates for different groups of dinosaurs with different foot and leg lengths. Whereas Alexander had used four times footprint length to estimate hip height or leg length of any dinosaur, Thulborn showed that, for accuracy, it is necessary to use a different leg length measurement for each different dinosaur group (see Fig. 6.2). He showed how we can derive the appropriate length from actual measurements taken directly from dinosaur skeletons. He concluded that hip height was closer to about four-and-a-half times foot length in small bipedal dinosaurs and five-and-a-half times foot length in large bipedal dinosaurs.

Thulborn further elaborated his studies to suggest that dinosaurs, like various living animals, adopted distinct "pre-

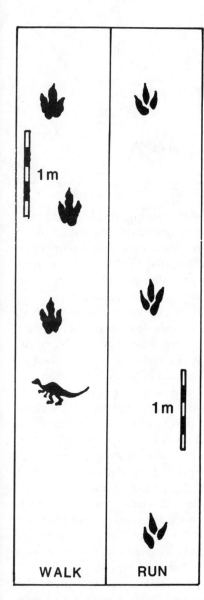

Figure 6.1. Complete strides (left-right-left) for medium-size theropod dinosaurs, showing different patterns for walking and running gaits.

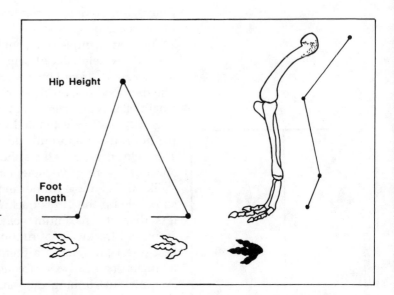

Figure 6.2. *Estimating dinosaur hip height. The simplest method is to multiply footprint length by four. A more accurate method involves using the precise ratio of foot to limb length, which is derived from the study of actual skeletons.*

ferred gaits," which he defined as walking, trotting, and running. Based on a survey of a large number of bipedal dinosaur trackways he found that few trackways indicate the intermediate, trotting gait. He suggested that this could mean that dinosaurs switched quickly from walking to running because the trotting gait was tiring or energetically inefficient.

After Thulborn's substantial contributions to this field, Robert Bakker, well-known advocate of agile dinosaurs, entered the debate with a catchy but misleading title "The Return of the Dancing Dinosaur."[5] In this 1987 contribution he claimed to have spent the previous three years engaged in a "comprehensive study of tracks . . . to compare dinosaur speed with that of land vertebrates from the Carboniferous to the present." He was evidently unaware of the work done by Thulborn and so devoted part of his study to reinventing the wheel and explaining how dinosaur limb length varied. One of his conclusions was that limb length is between five-and-a-half to eight times foot width in ornithopod dinosaurs.

He also introduced the concept of "cruising speed," which would generally be seen in the fossil record, rather than top speed, and further asserted that "dinosaurs had cruising speeds as high or higher than mammals with comparable body size and feeding habits" and that "theropod dinosaurs cruised at higher speeds than modern mammals." All this gives the reader the impression that dinosaurs were constantly moving at high speed. Cruising technically means moving at the most efficient speed for sustained travel over long distances, so it could simply mean walking. However, in today's high-speed mechanized world, and particularly in the context of the above statements, it is easy to interpret cruising as sustained *high-speed* activity. What Bakker appears to be referring to is simply "walking," the normal activity represented by 99 percent or more of all known dinosaur trackways. He was not referring to trotting, running, or any other fast or unusual gait. His conclusion, that "the average everyday pace of the dinosaur locomotor activity was as quick or quicker than that of the present day Mammalia," must therefore be taken with a substantial pinch of salt. The trackways do not prove this.

Thulborn's latest words on the subject interject a more moderate tone to the debate and provide a less fanciful view of "dancing dinosaurs." He suggests that only small cursorial dinosaurs were capable of running at high speeds and actually leaving the ground in an unsupported or suspended phase (i.e., "when the animal lifts all feet clear of the ground and 'floats' through the air under its own inertia").[6] He points out that large mammals like rhinos and elephants (three to six tons) do not attain a full gallop, with all feet off the ground. As awesome as their charges might be, according to Thulborn, they technically do not exceed a fast trotting gait. Gregory Paul disputes Thulborn's more cautious view by reminding us of the galloping white rhino that starred in the movie *Hatari*.[7]

Bakker may have done us a service in asking us to believe in his conjectures about acrobatic, dancing, even pirouetting dinosaurs. Certainly he has encouraged us to stretch our imaginations to the limit and forced us to think of how active and agile dinosaurs may have been. Had we ever witnessed a charging *Triceratops*, its fast trot would seem terrifying enough, and in retelling the story the charge could easily be embellished into a record-breaking gallop. But if Thulborn is right, a cool-headed, fleet-footed human might just have been able to outrun, or at least outmaneuver, any of the large quadrupedal dinosaurs. Africans tell us one can outrun a hippo, especially with a few well-timed changes of direction.

Using Alexander's formula, the running hippo trackway in Chapter 2 suggests speed of about 5 meters per second, or 100 meters in 20 seconds. Having once run a 100 meters in about 11.5 seconds, I still fancy my chances at zipping in at less than 20 seconds, even on a firm mudflat. But no challenges, please!*

*Sometime after writing this, I was astonished to read an article in which Alexander claims: "I am probably fast enough to outrun a pursuing tyrannosaur, but perhaps fortunately, I am unlikely to try."[8] Readers may indeed wish that our brash claims could be put to the test!

Any irregularity in a dinosaur's step and stride pattern usually indicates a change in behavior. A progressive increase in step length indicates that the trackmaker is speeding up or accelerating, whereas decreasing step length indicates slowing or deceleration. Other irregularities we sometimes encounter include various curves or swerves in otherwise straight trackways. Usually little can be read into such minor changes

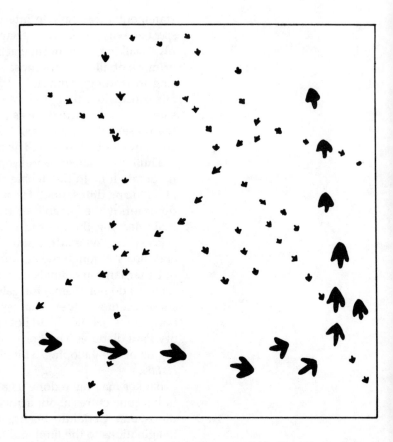

Figure 6.3. Trackway of a carnivorous dinosaur making a sudden right turn. *Lower Cretaceous of Canada (after Currie 1989).*

of direction. Footprints sometimes depict slippage on a soft substrate, and we can imagine a slightly off-balance dinosaur contending with slippery conditions underfoot.

Two unusual examples of dinosaurs making abrupt turns are known. The first trackway shows where a larger Cretaceous carnivore abruptly turned to the right,[9] and the second shows where a Jurassic brontosaur also veered quite sharply to its right (see Figs. 6.3 and 6.4).[10] We can only speculate about the reasons that prompted these individuals to change direction so suddenly (Chapter 13). Some threat or distraction nearby, or perhaps they simply changed their minds. In each of these cases we know only that there were other dinosaurs in the area at about the same time.

One of the most interesting and endearing examples of dinosaur behavior was reported by Roland T. Bird, in his interpretation of an *Anomoepus*-like (ornithopod) trackway in Edward Hitchcock's famous collection at Amherst College.[11] The trackway simply indicates an animal walking along, crouching or squatting down, then rising and proceeding again. In this exceptional case, however, the reason for this pattern of behavior is evidently that the animal sought refuge from a rain storm. The tracks leading up to the crouching or resting marks, like all the surrounding sediment, are pock marked with rain-drop impressions. Where the animal crouched down, however, there are no rain-drop marks, indicating that it remained crouched until the downpour abated. Then, when it proceeded after the squall had passed, it obliterated the freshly made rain pits as it moved off.

Occasionally individual dinosaurs were wounded or diseased, and in some cases the result was that they could not walk properly. Shinobou Ishigaki, a Japanese tracker in North Africa, encountered a unique trackway of a limping dinosaur.[12] This particular animal was afflicted in such a way that the fourth toe on its right foot did not stick out normally. It also took irregular short and long steps as a result of its affliction (Fig. 6.5). Another example, reported by Abel[13], is of a Jurassic trackway from New England made by a dinosaur that had apparently lost a toe. The trackway is of the *Eubrontes* type and clearly shows that the inside toe (digit II) of the right foot is missing (Fig. 6.5).

Here it is worth pointing out that measurements of trackways often indicate alternating long and short steps (the length differences are slight). We do not really know why this is. Possibly dinosaurs adjusted their steps slightly if they were turning to look sideways. More pronounced differences in long and short step lengths probably indicate changes from a normal walk to a trot. Such obvious changes are commonly observed in the trackways of modern four-footed animals.

The dinosaur footprint literature contains many examples of unusual behavior inferred from dinosaur and other reptile trackway evidence. Unfortunately many of these inferences are highly speculative, and it is necessary to devote a subsequent chapter to a reevaluation of some of the more bizarre and controversial interpretations (Chapter 13). However, we can discuss one well-known debate here. The question is, can dinosaur track evidence be used to demonstrate whether dinosaurs were capable of swimming? Obviously a dinosaur swimming out of its depth leaves no tracks, whereas one walking in very shallow water leaves a trackway that resembles normal walking progression. But what about the intermediate situation, where the dinosaur is partially buoyant? Consider what kind of tracks a dinosaur might leave if it were paddling along almost out of its depth, touching the bottom only sporadically, without impressing its full weight.

In 1944 Bird reported that he had found such evidence. From a trackway consisting almost entirely of sauropod front-foot impressions, he inferred a swimming brontosaur (see Fig. 6.6).[14] It apparently had touched the bottom only once with its left rear foot as it changed direction, veering to the right.

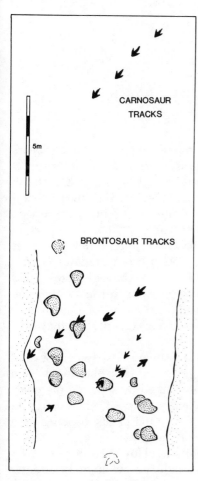

CARNOSAUR TRACKS

5m

BRONTOSAUR TRACKS

Figure 6.4. Trackway of a brontosaur making a right turn, Late Jurassic of Utah.

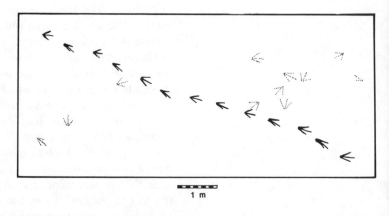

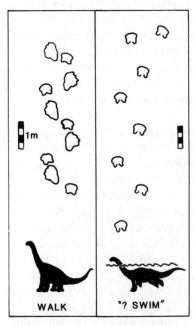

Figure 6.5. Top: *Trackway of a wounded bipedal dinosaur, discovered by Shinobu Ishigaki. Bottom: Trackway of a dinosaur with a missing toe (after Abel). Both examples are Jurassic.*

1 m

TRACKWAY OF A DINOSAUR

WITH A MISSING TOE

WALK   "? SWIM"

Figure 6.6. *Comparison of normal walking trackway of a brontosaur and supposed swim tracks. Note that the "swim" tracks have also been interpreted as underprints.*

Based on the evidence, such an interpretation at first appears reasonable. We need to try to explain incomplete or partial trackways, and it would be ludicrous to suggest that the brontosaur walked only on its front feet. Bird's interpretation was supported by the then-prevailing notion that sauropods were aquatic or semiaquatic and barely capable of supporting their own weight on land. It was also thought that, like other large herbivores, they sought refuge from predatory carnivores by fleeing into the water, where the carnivores could not pursue them. Such an interpretation carries with it the assumption that carnivores could not swim or that they at least perceived themselves to be at a disadvantage if they pursued prey into the water.

Bird's "swimming sauropod scenario" spawned several dozen papers on swimming dinosaurs. His conclusions have been both supported and undermined by separate discoveries in the 1980s and 1990s. The discovery of other sauropod trackways consisting mainly of front footprints, by Ishigaki in Morocco,[15] appears to suggest more evidence of partially buoyant or swimming sauropods. However, a report by Coombs of carnosaur tracks made by an animal paddling through a lake suggests that predatory dinosaurs could swim; the water was not necessarily a safe refuge for herbivores.[16]

Of course, evidence of a swimming carnivore does not in

itself discredit Bird's interpretation. There are, however, other possible interpretations that must be considered. For example, four-footed animals can leave front and hind footprints of different depths. The Texan and Moroccan front footprints are very shallow and occur in thinly layered sediment that we would expect to be severely depressed by the tracks of large animals. This suggests that the forefoot tracks could simply be undertracks associated with hind footprints that are less deeply impressed in the underlayers.

As explained in Chapter 3, understanding footprint preservation is very important if we want to avoid misinterpretation of the track record. Because of prevailing opinion in Bird's generation, he had quickly inferred that the tracks were those of a swimming animal. Brontosaur trackways had only just been discovered, so there was little direct evidence that they could walk on dry land. Consequently Bird was asking a pertinent question for the time, "Did *Brontosaurus* ever walk on land?" Now, a half century later, we know from many trackways that *Brontosaurus* could walk on land, and we are asking the opposite question "Did *Brontosaurus* ever swim out to sea?"[17] If the undertracks interpretation of front footprint–dominated trackways is right, the answer is that the footprint evidence does *not* support the swimming scenario.

As discussed further in Chapter 13, Jeff Pittman recently discovered that Bird had overlooked a hind footprint in this famous sequence. This discovery further weakens the swimming scenario. All this may not make for such an interesting interpretation, but it is better science.

Did dinosaurs drag their tails? This question about individual dinosaur behavior is frequently asked, and we recognize that trackways provide crucial evidence to help us formulate answers. Either a tail drag mark is present – as in a lizard trackway – or such traces are absent – as in most mammal trackways. Almost invariably dinosaur tail drag marks are absent. This indicates that they walked erect, as depicted in most modern reconstructions, rather than slithering or sprawling close to the ground or progressing with drooping tails, as many older interpretations suggest. As discussed in Chapter 13, it is tough on the tail to drag it, and complicated explanations of the lack of tail drag marks are unnecessary. The rarity of tail traces can simply be taken to mean that dinosaurs did not make a habit of dragging their tails.

Conclusion

Tracks provide abundant evidence of day-to-day activity of walking, but the evidence of unusual behavior is rare. There are some authentic examples of running, turning, and limping dinosaurs, but these are exceptional. The use of tracks to infer swimming behavior is fraught with problems and has led to flawed scientific literature and unsubstantiated assertions. Moreover, trackway evidence suggests that dinosaurs did not drag their tails. We shall return to some of these myths and misconceptions in due course when we examine how a purported hopping dinosaur trackway turned out to be the trail of an entirely different animal, and how other reports of outlandish activity often have simpler and more mundane explanations (Chapter 13).

The message is this. Tracking dinosaurs and interpreting their behavior is science as well as fun. After we have described the trackways, our interpretations should be based on the evidence. There may be more than one possible interpretation of a set of tracks, but that is okay: Science accepts multiple working hypotheses and unanswered questions. What it does not accept is promotion of a single dramatic hypothesis while other likely interpretations are ignored.

7

Social behavior

Large carnosaur amidst small theropods and ornithopods; based on published interpretations of the Cretaceous Lark Quarry site, Australia. Artwork by Edward Von Mueller.

Other trail disclosures indicate that no less than a dozen sauropods had crossed this section. All were progressing in the same direction as a herd. . .
  – Roland T. Bird, *"A Dinosaur Walks into the Museum"*
                                                    (1941)

Although many animals, from insects to fish, birds, and mammals, are gregarious to some degree, reptiles, including dinosaurs, were traditionally *not* regarded as social animals. However, in recent years we have witnessed interesting discussion about social behavior among dinosaurs. Some has been speculative, but much has been firmly based on exciting new paleontological evidence. Much of the evidence comes from the realm of trace fossils, either of large nesting colonies or of tracksites with multiple trackways, attributable to large herds. (Nest site evidence is outside the scope of this book, but such evidence is generally relevant to the social behavior debate. Because some dinosaurs, like many modern bird species, congregated to nest, they at least began life in a social or gregarious setting. Even if they afterward dispersed, they subsequently regathered during breeding season, presumably on an annual basis.)

Recent discoveries of multiple trackway sites indicate that many dinosaur groups were habitually gregarious. In 1941 Roland T. Bird was the first to document a site where twelve sauropod trackways trended in exactly the same direction, evidently indicating the passage of a herd. Subsequently he presented evidence for a herd of at least twenty-three brontosaurs (Fig. 7.1).[1] This latter discovery from a site known as Davenport Ranch has become famous in the social behavior debate. After Bird's initial report, there was very little discussion of the implications of such trackway evidence until the Dinosaur Renaissance of recent years. In 1972 John Ostrom published a paper entitled "Were Some Dinosaurs Gregarious?" in which he examined trackway evidence from several sites, including Davenport Ranch.[2] He demonstrated that several other tracksites yield evidence of multiple trackways trending in the same direction or with a "preferred orientation." This preferred orientation or preferred directional trend appeared to characterize small Jurassic theropod trackways from Mount Tom Park, Connecticut, and larger theropods from the Cretaceous of Texas.[3] When we include the sauropod evidence, we find that many very different dinosaur groups may have been gregarious. Since the publication of Ostrom's paper many more new sites have displayed evidence of gregar-

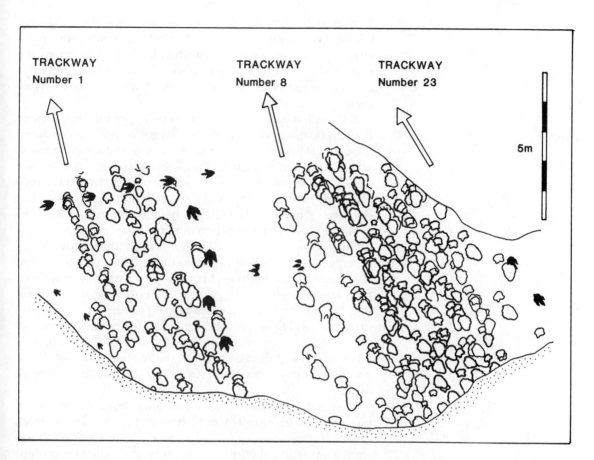

TRACKWAY
Number 1

TRACKWAY
Number 8

TRACKWAY
Number 23

5m

Figure 7.1. Parallel trackways of
brontosaurs at Davenport Ranch,
Texas, provide convincing evidence for
a herd of at least twenty-three animals,
all moving in the same direction.
About five theropods also crossed the
area in various directions. Brontosaur
trackways are numbered from left to
right. (Redrawn after Bird, 1944.)

iousness or herding. These include sauropod trackway sites
in Africa and North and South America, large ornithopod
trackway sites in North America and Korea, a small ornithopod site in Spain, large theropod trackway sites in South
America, and a small theropod–small ornithopod trackway
site in Australia (see Chapter 8). The evidence, in fact, suggests that *most* dinosaurs were gregarious.

Before discussing some of these sites in detail, we should
review the criteria used in identifying trackways of herds. Ostrom was careful to point out that the presence of many trackways trending in a preferred direction does not necessarily
constitute proof of gregariousness. It is always possible that
a series of individual animals could have passed on separate
occasions, choosing the same direction because of what he
called some "physically controlled" feature of the landscape,
such as we might find today in a narrow river valley or along
a shoreline or river bank. If there is only one pathway along
which animals can pass, then the orientation of their trackways is strictly predetermined. They can only pass in a given
direction, from A to B, or in the opposite direction returning

from B to A. Ostrom suggested various ways of distinguishing herd evidence from physically controlled pathways in the landscape. For example, if individual trackways cut across the preferred trend there is no evidence for controlled pathways and the evidence for gregariousness is much more convincing.

Although the evidence at most sites allows for some leeway in interpretation, recent work has suggested that we can adopt a more systematic approach to the analysis of multiple trackways. First, we must note that Ostrom's suggestion of two possible interpretations is somewhat incomplete and artificial. Evidence of a physically controlled pathway does not disprove a herding hypothesis, because landscape configurations influence the direction of herds in much the way they do individuals. As more preferred orientation trackway sites emerge, so, too, does evidence of their relationship to ancient landscape configurations. Many herds appear to have been following shorelines, suggesting that both Ostrom's controlling factors worked in combination to produce the trackway orientations. These topics are analyzed further in Chapter 10. Other factors, such as seasonal or even daily migration, may have influenced dinosaur trackway directions, though these are difficult if not impossible to demonstrate with fossil track evidence.

We need to consider at least two other lines of evidence in our analysis of probable herd trackways. The first is track depth. If the tracks of adjacent animals of comparable size are similar in depth, then the consistency and moisture content of the sediment must have been similar at the time the animals passed. If we find that track depth varies along a trackway, it is useful to determine if a similar sequence of depth changes occurs in adjacent trackways. If the changes are consistent from trackway to trackway, this will also suggest that the animals passed at about the same time. By contrast, parallel trackways of different depths are likely to suggest animals passing on separate occasions, perhaps days, weeks, months or even years apart.

A second very interesting line of evidence relates to "intertrackway" spacing, the lateral space between adjacent trackways. In recent years a large number of parallel or sub-parallel trackways have been reported showing very regular patterns of intertrackway spacing. This evidence, which suggests animals walking shoulder to shoulder or in some kind of regular formation, appears to argue strongly in favor of herding. Such regular formations are to be expected and are reminiscent of formations of birds in flight or of soldiers marching.[4] It is a common observation that some birds fly in very regular formations. Studies of gregarious mammals like the African wildebeest also reveal regularly spaced trackways after the passage of herds or groups. In a group, animals instinctively prefer to allow a minimal space between individuals. One of the best examples was reported by Phil Currie, from a Cretaceous tracksite in western Canada, where he found evidence for a herd of about a dozen *Iguanodon*-like ornithopods, all walking about one to two meters apart.[5] At one point five of the trackways swerve right before resuming a direct course (Fig. 7.2). This appears to suggest animals walking on a broad front. When one began to veer out of line several others swerved in unison to avoid collision. There are many other examples of parallel, equally spaced ornithopod trackways, including a layer in the Cretaceous Jindong Formation of Korea, where at least twenty *Iguanodon*-like individuals were also walking on a broad front in this highly regular fashion.

In addition to the Canadian and Korean evidence for regular spacing of individuals in ornithopod herds, similar evidence exists for herds of various other groups, including brontosaurs and theropods. One of the better brontosaur examples comes from Dinosaur State Park along the Paluxy River in Texas (Figure 7.3) and differs significantly from the Davenport Ranch example. As explained in detail below, the Paluxy site indicates at least twelve large sauropods walking in the same direction at regularly spaced intervals and followed later by at least three theropods. The trackways from Dinosaur State Park have often been taken as evidence of a single brontosaur being followed or attacked by a carnivore. As explained in Chapter 13, this is not the case, and Roland T. Bird knew it: The evidence suggests a herd of brontosaurs followed later by at least three carnivores.[6] Although Bird never mapped all twelve trackways, he did map segments of at least five trackways (see Fig. 7.3). Unlike the Davenport Ranch site, where twenty-three individuals passed over a very narrow area only 15 meters across, the Paluxy herd was more widely dispersed.

The different trackway patterns are clearly easy to distinguish. The Paluxy sauropod trackways never cross, whereas the Davenport Ranch trackways show considerable overlap, at first sight appearing to be a jumbled mess of footprints. When carefully analyzed, the Davenport Ranch trackways clearly indicate that the twenty-three animals crossed the area in a definable sequence, with a considerable amount of following in line[7] (see Chapter 13).

The reasons why some herds progressed on a broad front whereas others were more strung out in line is not known. It may have had something to do with animal size, herd size, or available space in a given area. The Paluxy sauropods were very much larger than the Davenport trackmakers, so pre-

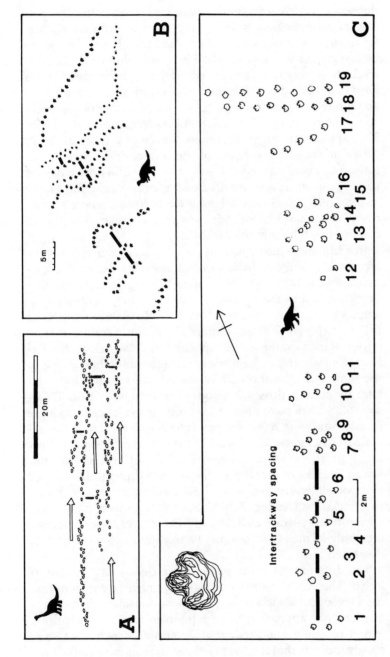

Figure 7.2. Parallel, regularly spaced trackways are strong indicators of dinosaur herd activity. A: Brontosaurs from the Jurassic of Colorado; B: ornithopods from the Cretaceous of Canada; C: ornithopods from the Cretaceous of Korea.

B

5 m

A

20m

C

Intertrackway spacing

2m

1 2 3 4 5 6 7 8 9 10 11 12 13 14 15 16 17 18 19

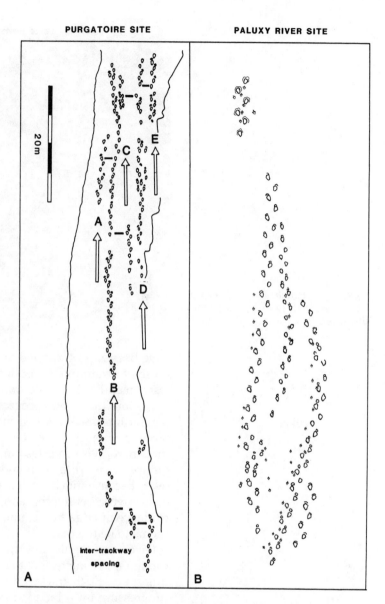

PURGATOIRE SITE   PALUXY RIVER SITE

Figure 7.3. Parallel trackways of brontosaurs from the Jurassic of Colorado and the Cretaceous of Texas.

sumably they needed more space when progressing as a herd. However, the evidence of five small sauropod trackways at the Late Jurassic Purgatoire site in Colorado (Fig. 7.3 and 7.4) shows that smaller sauropods sometimes spread out and progressed on a broad front without leaving overlapping trackways.[8] It appears reasonable to speculate that herds got congested when they were large or where space was limited. Smaller herds with available space could spread out.

There is not much well-documented evidence for regular intertrackway spacing in multiple carnivore trackway assemblages. This could be because some carnivore trackway sites

Figure 7.4. *Parallel trackways of brontosaurs, Late Jurassic, Purgatoire site, Colorado.*

have been misinterpreted, and the footprints confused with those of ornithopods.[9] Both the Mount Tom site reported by Ostrom and another site from South America reveal evidence that some theropods progressed in herds. If we accept that regular intertrackway spacing is less common among carnivorous dinosaurs, then this could be taken as an indication that theropods were less inclined to progress in structured groups than were herbivores, their probable prey. (Moreover, because much evidence points to a low predator/prey, or carnivore/herbivore, ratio,[10] we might expect to find more evidence of large herds among the herbivores.) To date, this seems to be the case.

Given that regular intertrackway spacing patterns are quite common, it is interesting to speculate on herd structure among dinosaurs. Such trackway evidence either suggests animals progressing on a broad front or in a spearhead, *V*-shaped, formation as sometimes adopted by flying birds. We know that gregarious animals ranging from fish to mammals are conscious of space allocation within the herd. When in motion they may appear to move in unison like a superorganism. There is no reason for us not to suppose that this was also the case with some dinosaurs. The trackway evidence quite strongly supports such a conclusion.

The term *herd structure* is very loosely defined. As outlined in the preceding paragraph it essentially means a discernible pattern. The term was first introduced in 1968 by Robert Bakker when he suggested that the Davenport Ranch trackways indicate that large brontosaurs traveled on the flanks of the herd protecting juveniles in the center.[11] This appealing idea fits in well with the social dinosaur scenario but is only an educated, off-target speculation; there is no trackway evidence for it. As we shall see in Chapter 13, the footprint evidence tells a different and much more complex story when we take the time to decipher it carefully.

In addition to the evidence for gregarious herds of dinosaurs presumably belonging to a single species, there are a few sites that suggest interaction between species. There are at least two examples of carnivores following sauropods. The first example, from the Paluxy site, has been mentioned on several occasions. Frequently misrepresented as an example of a single carnivore attacking a single sauropod, the evidence clearly suggests at least three carnosaurs following twelve or more sauropods (see Chapter 13). An even more spectacular example of this type was recently reported from the Late Cretaceous of Bolivia, where at least fifty theropods were following a group of sauropods.[12] In both cases the overlap of theropod on sauropod trackways proves that they followed soon afterward, at a time when the sediment consistency was similar. Footprint evidence suggests that there was an important ecological predator–prey relationship between sauropods and large theropods from the Jurassic through the Cretaceous (see Chapter 8). Given this fact, the trackway evidence can reasonably be interpreted as stalking, the dynamic expression of the predator–prey relationship.

The last well-documented example of a footprint site yielding evidence of social behavior and species interactions comes from a remote site called Lark Quarry in Queensland, Australia. The site was studied by Tony Thulborn and Mary Wade. They uncovered several thousand very small footprints, all with a strong preferred orientation. They also documented a single trackway of a large biped of theropod affinity. Because the small bipedal tracks were all interpreted as those of running individuals, the trackways were interpreted as evidence of a "dinosaur stampede."[13] The theropod was assumed to have instilled panic among the flock of diminutive bipeds, which fled for their lives as the threatening predator drew near – more tooth and claw drama in the Mesozoic. Greg Paul has bluntly disputed the Thulborn and Wade stampede story by claiming that the large dinosaur was an ornithopod, not a carnivore. At any rate, although it is hard to prove that the large trackmaker was the villain, the small dinosaurs were certainly running in one direction and a large dinosaur was present at the site.[14]

The site is of considerable interest for another reason. According to Thulborn and Wade, careful analysis of the abundant small trackways indicates that they may be divided into two distinct types, representing small ornithopods and small theropods in almost equal proportions. This suggests the mixing of two herds, flocks, or gregarious groups, either during normal daily activity before the carnivore approached or during the crisis of panic that ensued when the predator threatened.

Such mixing of herbivorous ornithopods and carnivorous theropods in a single herd, at first sight seems strange, and the interpretation may be incorrect. The case may be, as Greg Paul suggests, that the theropods were simply chasing the ornithopods. Alternatively, the trackways may have been misinterpreted: They might all be trackways of theropods! However, a less aggressive scenario is still a possibility, especially if the small theropods did not threaten the ornithopods by preying on them. Evidence from nest sites has shown that different species sometimes nested together, presumably sharing the responsibility for protecting the colony. The two small Australian trackmaking species may conceivably have confused and foiled the large predator more effectively by fleeing together. Deliberately or inadvertently they could have created a more thorough distraction.

Activity levels

The number of tracks left by an individual depends on several important factors, the first being how often it frequents or steps on soft receptive ground where tracks will be formed. An equally important factor is the individual's activity level. We speak here of the behavior of various species, not of individual traits. Some species are more active than others. As we know from observing living mammals and birds, it is very often the small species that are most active. A mouse or squirrel scurries around in almost frenzied activity compared to the slow, measured movements of an elephant. This does not mean that the mouse may not sometimes stop or rest, or that the elephant never gets galvanized into more energetic activity. What is important is that the mouse is much more active than the elephant, on average. If both species were making tracks all the time, we would expect to find dozens, even hundreds of small mouse tracks for every large elephant track. (As we know, however, this is not always the case. Elephants are heavier and their tracks are larger and deeper, covering and disturbing much more ground. They are therefore much more obvious, and likely to be preserved on substrates where mice would fail to leave any trace of their activities.) Because few dinosaurs were as small as mice, we need a better analogy. Many claim that birds are the dinosaurs' closest living relatives, and like mammals they vary considerably in their activity levels and patterns. Shore birds, such as large herons, are fairly slow and measured in their movements; it may take them several minutes, even hours to make a few tracks. By comparison small waders are frequently on the move, often engaging in bursts of rapid activity in which they make hundreds of tracks in a matter of a few minutes. The tracks of small waders will probably outnumber those of larger birds like herons. Because the disparity in size is less than that between mice and elephants, we might expect the small tracks to also be preserved in greater abundance. This is what we see in the fossil record from Cretaceous times onward. Tracks of waders and shore birds are now known to be very abundant (see Chapter 9).

As already discussed, dinosaurs, particularly small theropods, were very birdlike, and their adaptations indicate well-developed running ability. It is therefore reasonable to assume that chicken- and turkey-sized species were much more active than a large *Tyrannosaurus rex* or a giant brontosaur. These larger animals had to overcome considerable inertia to get moving.

As we shall see in the next chapter, activity levels could have a significant bearing on estimates of dinosaur population size made using tracks. We now have enough data on footprints to begin to use them in census studies of dinosaur communities. Our interpretations must, however, take the activity levels of different species into consideration. For example, theropod tracks are often much more abundant than their bones. This suggests that they probably left more tracks due to higher average activity levels.[15]

We will also have to take into account the fact that small active species may be represented by a larger number of individuals than larger, less active species. They may also be more social and gregarious. This obviously means that they can potentially make even more tracks than solitary species, even without the activity-level differences. The aforementioned Lark Quarry site needs to be considered in the light of these observations. We can clearly see that the small trackmakers were active (running) and relatively abundant, whereas the large individual was solitary and walking.

As we have seen, many sites reveal evidence of two or more different species or groups interacting together in various ancient environments. When we look beyond the evidence for or against gregarious behavior, we come up with other useful ecological evidence. This includes data on the number of trackways made by different species, which is in part controlled by species' activity levels and size and hence their potential for preservation in the fossil record. Once these factors have been taken into consideration, we can look at how the various numbers or proportions of different track types provide useful census counts. In effect, we can use footprints to determine the proportion of different trackmaking groups represented at various sites. Such census data give us insights into the composition of the dinosaur communities that existed in these areas. This is the subject of the next chapter – how tracks help reveal the ancient ecology of dinosaur communities.

Plate 1. *Top, left: A small Triassic dinosaur footprint preserved as a natural impression. Popo Agie Formation, Colorado. Top, right: A large Jurassic dinosaur footprint as it was found in the field, preserved as a natural cast. Entrada Formation, Utah. Bottom: A typical tracksite in a dry stream bed. Jurassic Wingate Formation, Cactus Park, Colorado.*

Plate 2. *Top: Tracks cover the surface of a house-sized block fallen from the high cliffs of the Green River Canyon, Utah. Jurassic Wingate Formation. Bottom: Multiple track-bearing layers exposed on the foreshore. Cretaceous Jindong Formation, South Korea.*

Plate 3. Top: *Aerial view of the Purgatoire Valley tracksite, Late Jurassic Colorado. Note several track-bearing layers. Bottom: Digging for tracks below the ice. Field work in Purgatoire in January.*

Plate 4. *Top, left: Trackway of a Jurassic bipedal carnivore. Entrada Formation, Utah. Top, right: Trackway of a Jurassic quadrupedal brontosaur. Morrison Formation, Colorado. Bottom: Trackers study the trackway of a turning brontosaur. Morrison Formation, Utah.*

Plate 5. *Top: BBC field crew shoots sequence on brontosaur locomotion. Purgatoire Valley, Colorado. Bottom: Excavating a Cretaceous duck-bill dinosaur track in a coal mine.*

Plate 6.   Top, left: *Parallel trackways of Jurassic theropods. Entrada Formation, Utah. Top, right: Trackers at a multiple trackway site. Jurassic Entrada Formation, Utah. Bottom: Five parallel brontosaur trackways, Morrison Formation, Colorado.*

Plate 7. Top: Trackers examine a giant brontosaur trackway at a Cretaceous site in South Korea. Note that tracks are underprints. Small tracks also visible on a lower level. Bottom, left: A trampled surface on a fallen block of Jurassic sandstone, Utah. Bottom, right: A trampled surface in the Jurassic Entrada Formation, Utah.

Plate 8. *Top: A tracker ponders the future of a well-preserved track. Note trampled layers in the background. Jurassic Navajo Formation, Utah. Bottom, left: Trackers excavate a large Cretaceous sandstone slab that reveals dinosaur and bird tracks. Cretaceous Dakota Group, Colorado. Bottom, right: Well-preserved Cretaceous tracks are preserved in a well-designed shelter. La Rioja Province, Spain.*

8

Ancient ecology

Ceratopsian trackmaker, based on late Cretaceous footprint evidence from Colorado. Artwork by Edward Von Mueller.

The fauna as shown by the tracks alone must have been un-usually full.
  – Edward Hitchcock, *Ichnology of New England* (1858)

In the previous chapter we learned how tracks can provide insights into species interactions. However, social behavior and predator/prey confrontations are not the only aspects of ancient ecology that can be studied using tracks. Like modern animal communities, ancient dinosaur communities were complex entities, components of larger ecosystems that included plants, invertebrates, and other nondinosaurian vertebrates. In recent years much has been written about ancient ecology, or paleoecology, and the Mesozoic ecosystems inhabited by dinosaurs.[1] Little is known, however, of the considerable contribution that footprints can make to this field of study.

It is worth recalling that tracks are a dynamic census of once-living animals. Unlike bones, which may be fragmentary and transported some distance from an animal's original habitat after its death, tracks always represent living animals engaged in activity in some part of their original range. Although it has not usually been appreciated by paleontologists, at least until recently, tracks are often far more abundant than bones. A given formation is considered moderately productive if it has yielded the partial remains of only a few dozen dinosaurs belonging to a handful of species. The extent to which such skeletal remains contribute a useful or reliable census of the dinosaur community is open for discussion. At any rate, bones have always been used as the main evidence for reconstructing dinosaur communities. The practice of using tracks is new and may be considered radical by some.[2]

A given formation, however, may yield the trackways of a large number of individual dinosaurs, including some that are clearly not represented among the skeletal remains. In such cases the track evidence adds significantly to what is known of the dinosaurian community and may be statistically more reliable than the body fossil record. In extreme cases, where a formation has yielded little or no bone material, but an abundance of tracks, the footprints may provide an otherwise unavailable census. A good example is the Cretaceous Jindong Formation of Korea, which has yielded the trackways of several hundred individual dinosaurs but not a single bone. In fact, there are virtually no dinosaur skeletal remains known from the entire Republic of Korea, yet its track record is spectacular. Parts of the Cretaceous Dakota Group of Colorado provide a similar example. Tracks are so abundant in the Dakota Group that we have coined the term "Dinosaur freeway" to describe the extensive layers of track-bearing strata.

The use of trackways in census paleoecology is essentially a brand-new field, and as such it has yet to stand the test of time. Trackways have the advantage of being the in-habitat evidence of dinosaurs, but there are also some disadvantages. The most serious drawback is that track types cannot be assigned to particular species; a census list may simply read 50 percent theropod, 40 percent sauropod, and 10 percent ornithopod, although in some cases it is possible to define several distinctive track types within these broader categories.[3]

Traditionally tracks have been regarded as useful only in areas where little or no other fossil evidence exists. A much better approach is to view tracks as an integral part of the entire paleontological evidence. In a given deposit they may constitute all the evidence, only a part, or none. Wherever they exist, they provide evidence that falls into one of the following categories[4]:

1. Footprints are the only evidence of dinosaurs.
2. Footprints constitute the majority of available evidence:
   a. Consistent with known skeletal remains
   b. Not consistent with known skeletal remains.
3. Footprints and bones occur in about equal proportions:
   a. Consistent with known skeletal remains.
   b. Not consistent with known skeletal remains.
4. Footprints constitute only a small portion of the available evidence:
   a. Consistent with known skeletal remains.
   b. Not consistent with known skeletal remains.
5. Footprints are unknown.

In this system paleontologists can rank track-bearing deposits on a scale from 1 to 4 for their relative importance in providing new paleontological information. In all cases, however, tracks are important, for they either provide new information that supports or is consistent with previous findings (situation a), or they provide supplemental evidence, at variance with the known record (situation b), therefore requiring us to reexamine the deposit.

Once we recognize and systematically define the contribution footprints make to our understanding of dinosaur communities, a more rigorous and credible scientific analysis is possible. We can characterize various tracksites and track-bearing deposits in one of the above categories and systematically evaluate the information yielded. For a coherent perspective we can take a trip through time and outline the contribution made to dinosaur paleoecology by tracksite analysis, beginning early in the Age of Dinosaurs (Late Triassic) and proceeding to the end (Late Cretaceous). To avoid undue bias in favor of the large number of recently studied track-dominated deposits (cases 1 and 2), we will also examine various well-known bone rich formations (cases 3 and 4).

Late Triassic Paleoecology

North America

A number of Late Triassic deposits are famous for diverse fossil vertebrate faunas dated early in the Age of Dinosaurs. Examples of such deposits include the Chinle Formation and its equivalents in the Western United States, famous for the petrified forest remains and painted desert scenery in the region of northeastern Arizona. Since dinosaurs were scarce at this time, it is not surprising their bones are relatively rare. One unusual site, the *Coelophysis* graveyard at Ghost Ranch in New Mexico, yields abundant skeletons.[5] Elsewhere, bone accumulations are dominated by the remains of various other reptiles, including the large crocodile-like phytosaurs, the armored herbivorous aetosaurs, bovine-like, mammal-like synapsids, and various other species. In the final analysis dinosaurs were only a small component of the fauna, comprising about 5 percent.

In the Chinle Formation and deposits of the same age, such as the Dockum Group of New Mexico and Texas and the Popo Agie Formation of Colorado and Wyoming, we now know of about twenty tracksites. Until recently they would have been considered type 4 deposits lacking much in the way of valuable track information. New tracksite discoveries have elevated several to ranks 2 and 3. At least three of the larger tracksites have yielded enough tracks to allow us to attempt statistically meaningful paleoecological censuses. Although each formation yields a different assemblage of tracks, they all show that dinosaurs were a very minor component of the trackmaking fauna. The few known dinosaurian tracks indicate the presence of small *Coelophysis*-like forms. In the case of the Chinle Formation, some larger tracks reveal the presence of another bipedal dinosaur, as yet unknown from skeletal remains. The tracks also tell us that armored herbivores known as aetosaurs were abundant leaving tracks known as *Brachychirotherium*. Thus we are able to reconstruct an animal community similar to that revealed by the study of skeletal remains, but with the added benefit of tantalizing glimpses of a few hitherto unknown larger dinosaurs in particular regions. The Chinle Formation therefore appears to be a type 3a deposit.

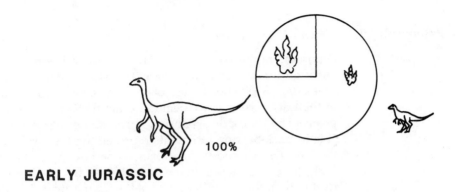

EARLY JURASSIC

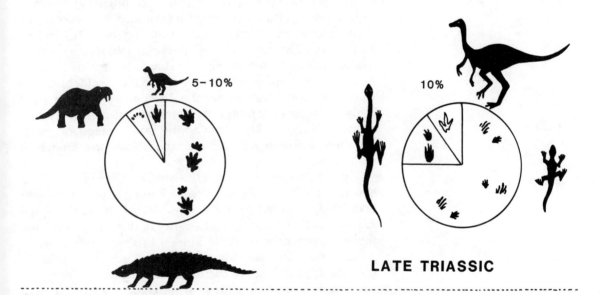

LATE TRIASSIC

Figure 8.1. Late Triassic and Early Jurassic dinosaur community censuses, based on trackway proportions. Triassic examples based on Chinle (left) and Popo Agie (right) footprint assemblages. Note the dramatic increase in the proportion of dinosaur tracks in the Jurassic.

In the case of the Popo Agie Formation, we find that a very different situation exists.[6] It had previously been known for a handful of large reptile fossils. When the track assemblage was studied, it was shown to indicate an entirely different suite of small reptiles, thus considerably expanding what was known of the ancient animal community. We can therefore elevate it from a type 4 deposit to as high as type 2b. The Popo Agie trackmakers appear to include a bizarre, long-necked, *Tanytrachelos*-like amphibious reptile, a Tuatara-like lizard, and a *Coelophysis*-like dinosaur, probably the oldest known from North America. As shown in the illustration (Fig. 8.1), the footprint census dramatically changes our view of life in Colorado and Wyoming during Popo Agie times. It is not that the bone census was wrong, just that it entirely missed all of the small animals represented by these tracks.

Early Jurassic Paleoecology

Eastern North America

Dinosaurs rapidly increased from a minority group to a majority group between latest Late Triassic and Early Jurassic times (Fig. 8.1). The success of the dinosaurs is clearly seen in the track record as we begin to examine Early Jurassic deposits. The dinosaur expansion is sometimes also recognizable in track-bearing sediments representing the very end of the Late Triassic epoch. In seeking a good example of sizable Early Jurassic dinosaur track assemblages we need look no further than the classic localities of the Connecticut Valley described by Edward Hitchcock in the early nineteenth century.[7] These footprints, which Hitchcock originally assigned to large fossil birds, have been reexamined by workers like Richard Swann Lull, and more recently by John Ostrom and Paul Olsen. The consensus of opinion is that they represent a dinosaur community dominated almost exclusively by small and medium-size bipedal theropods, which incidentally include coelurosaur-like forms, the direct ancestors of modern birds. Some of these footprint assemblages also include larger robust theropods like *Eubrontes*, the small semiquadrupedal ornithopod *Anomoepus*, and a crocodile-like form known as *Batrachopus*.

A good and easily accessible example of a site from this epoch is the Rocky Hill site, already known to us as Dinosaur State Park, Connecticut (see Appendix A). It occurs in the East Berlin Formation, one of a series of deposits comprising the thick sequence of strata known as the Newark Supergroup. Like many sites from this region Rocky Hill attests to the former abundance of theropods (*Eubrontes*, *Grallator*, and *Anchisauripus*) and constitutes what Ostrom terms "a most unusual community," one dominated by carnivorous forms. As discussed in Chapter 10, the Early Jurassic animals of the Connecticut Valley inhabited lake-filled Rift Valley environments, different from the setting in which many track assemblages are found.[8] However, the global evidence suggests that the dinosaur fauna was fairly typical of that found worldwide. Similar track assemblages are found in rocks of this epoch at widely scattered localities in Europe, Africa, and elsewhere in deposits that represent different ancient environments.

Because of the scarcity of skeletal remains in Lower Jurassic rocks of the Connecticut Valley region and the spectacular abundance of tracks, the footprint assemblages rank as 2b, high on the scale of relative paleontological importance. Many sites have yet to be systematically analyzed for the considerable volume of valuable statistical census information available. When this type of research is completed, our paleoeco-logical knowledge of the distribution of Lower Jurassic trackmakers in time and space promises to be considerably improved.

Western North America

In the Colorado Plateau region of northeastern Arizona, eastern Utah, and western Colorado, Lower Jurassic deposits are represented by formations like the Wingate, Moenave, Kayenta, and Navajo – distinctly Native American and western names. These formations differ dramatically from the rift valley deposits of the Newark Supergroup because they represent ancient Sahara-like sand seas. During the Early Jurassic we can visualize thousands of square kilometers of sand dunes with only the occasional small playa or oasis.

Traditionally these desert environments have been regarded as hostile to animal life. This interpretation has been strongly reinforced by the virtual lack of fossilized skeletal remains,[9] but the tracks tell a different story. Perhaps by now we will not be surprised to find that these deposits are replete with footprints (Fig. 8.2). This immediately elevates most of the formations from type 4 to type 2, or even type 1 deposits.

Let us choose as our example the Navajo Formation near Moab, Utah (see Appendix A). Here we find a number of sites that yield small and large theropod tracks (*Grallator* and *Eubrontes*), evidence that suggests trackmakers similar to those reported from eastern North America. However, we have also recently found *Otozoum* tracks and a large number of tracks attributable to mammal-like trackmakers and lizards.[10] This considerably expands the picture of the ancient animal community. *Otozoum* probably represents a prosauropod, like the track *Navahopus* (Chapter 5), but representing a larger, possibly bipedal species. We can therefore visualize a desert environment in which oases or watering holes were frequented by at least two types of carnivorous dinosaurs, probably at least two species of prosauropod, various mammal-like species, and lizards. The area was a living desert, not a wasteland.

Middle Jurassic Paleoecology

The Middle Jurassic epoch is poorly known from the viewpoint of dinosaurian paleontology. This is again largely due to the lack of documented skeletal remains, tracks, and eggs worldwide. This shortage of material sometimes complicates the geological process of assigning correct ages to rock sequences. In some areas we still do not know if certain rocks are really Middle Jurassic. Leaving aside technical rock-dating

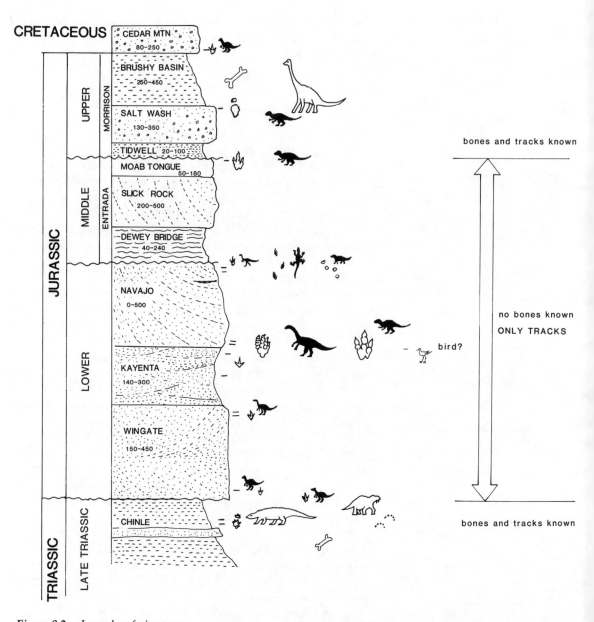

Figure 8.2. *Legend on facing page.*

problems, we can identify a few promising Middle Jurassic trackway assemblages that give paleontological insights into this epoch.

North America

The first, a remarkably extensive site in the Entrada Formation, also in the Moab area of eastern Utah, reveals evidence of significant developments in theropod evolution between Lower and Middle Jurassic times. The site is particularly spectacular for the sheer abundance of tracks and the lateral extent of the track-bearing layers. Although the site is still under investigation, it is possible to report on a series of remarkable footprint statistics. Tracks in the main track-bearing layer can be traced laterally for at least 15 to 20 kilometers along an almost continuously exposed bedding plane and less continuously due to sporadic outcrops over an area of at least 300 square kilometers. We call these large sites megatracksites (see Chapter 12). Individual sites that have been mapped indicate track densities from one to ten per square meter. Almost without exception, the tracks indicate medium-size to large theropods, with feet ranging from 25 to 30 centimeters up to 45 or 50 centimeters in length. This appears to represent another example of what Ostrom would call a "most unusual community."

Before jumping to the conclusion that such an abundance of tracks must represent a very large number of dinosaurs, in a desert environment of all places, it is worth stressing that high track densities may have geological explanations. As discussed in the Chapters 10 and 12, the track abundance is probably not attributable to exceptionally large dinosaur populations, but to repeated trackmaking activity in the same beds over a long period of time.

Paleoenvironmental considerations, outlined in Chapter 10, also help explain the carnivore abundance in desert environments. It is not surprising in fact, that deserts supported few herbivorous dinosaurs and that the food chain was dominated by carnivores. Tracks and trails from these Jurassic environments sometimes indicate the presence of various invertebrates, including scorpion-like forms, crayfish-like playa lake burrowers, and small *Coelophysis*-like, bipedal dinosaurs that may have fed on such small fry. The larger carnosaurs may have fed on small dinosaurs, mammals, and mammal-like species. There is no compelling reason to infer the presence of abundant herbivores in this particular dinosaur community. In fact, there appears to be a complete absence of the tracks of large brontosaurs. There is also no evidence of prosauropod dinosaurs surviving into the Middle Jurassic; nor have the tracks of any other herbivorous dinosaurs been identified.

When we interpret the evidence from the Middle Jurassic Entrada Formation, it is clear that the tracks represent an important type 1 site. The formation both in the Moab area and elsewhere in the region is essentially devoid of dinosaur remains attributable to these or any other trackmakers. Probably the best-known theropod from the area is *Allosaurus* from the younger, overlying Morrison Formation. Perhaps the Entrada tracks represent an *Allosaurus* ancestor of some kind, or

Figure 8.2. Tracks in early Mesozoic deposits of eastern Utah are important indicators of ancient animal communities. In ascending order, tracks indicate the presence of aetosaurs, therapsids, and dinosaurs (Chinle Formation); theropods and prosauropods (Wingate, Kayenta, and lower Navajo Formations); lizards, mammal-like reptiles, and small dinosaurs (upper Navajo); theropods (Entrada); and theropods and sauropods (Morrison and Cedar Mountain). In most of the sequence tracks are the only evidence of animal life.

a descendant of the *Eubrontes* trackmaker. In any event they fill an important gap of about 25 million years in the track record between the Early and Late Jurassic.

North Africa

The second interesting Middle Jurassic site represents a complex of fascinating trackway evidence from North Africa and sheds light on the rise of the sauropod dinosaurs. Traditionally the Late Jurassic has been known as the age of sauropods, or Age of Brontosaurs, even though it is known that true sauropods originated somewhat earlier. These Early and Middle Jurassic epochs are well represented by a sequence of limestone-rich strata outcropping in the rugged Atlas Mountains of North Africa, particularly in Morocco. As shown in Chapter 12, these limy deposits represent an ancient lagoon-strewn coastal plain, quite different from the desert sand dune environments that prevailed at this time in parts of North America. The environment was like the Florida Bay area or even the balmy lagoons of the Bahamas.

The first important dinosaur tracks were reported in 1937 from an area near Demnat, Morroco. However, it was not until 1978 that a reconnaissance exploration organized by the Moroccan Earth Sciences Museum discovered a considerable number of tracks, including those of large sauropods. Further exploration by Moroccan, French, and Japanese researchers revealed a large number of tracksites of Middle Jurassic age, as well as several dating to earlier epochs (the Lower Jurassic, or Liassic).[11] The footprint evidence reveals a region populated by dinosaurs from the Late Triassic onward, with a remarkable abundance of sauropods in Early and Middle Jurassic times. The early Jurassic footprints include the oldest-known true sauropod tracks, and the Middle Jurassic sites indicate that sauropods dominated the dinosaur communities in this area. If we remember that their tracks (and bones) are absent from the desert deposits of North America, we will appreciate just how important the environment must have been in controlling their distribution during this epoch.

The various North African formations have yielded dinosaur skeletal remains, including important sauropod material, but they are particularly rich in brontosaur footprints and should be placed in category 2 or 3. More significantly we have evidence that the North African dinosaur communities were appreciably different from those found in North America at this time. As we shall see in Chapter 10, this has much to do with how the ecology was influenced by differences in the ancient environment.

Late Jurassic Paleoecology

The Late Jurassic is sometimes known as the zenith of the Age of Brontosaurus. During this epoch sauropods like *Apatosaurus (Brontosaurus)*, *Brachiosaurus (Ultrasaurus)*, *Camarasaurus*, and *Diplodocus*, among others, reached their ascendancy. They shared their habitats with four-footed ornithischians like *Stegosaurus*, ornithopods like *Camptosaurus* (an iguanodontid), and theropods like *Allosaurus*.

By far the best-known and most fossiliferous dinosaur beds of this epoch are those of the Morrison Formation in the Rocky Mountain Region. Deposits assigned to this Formation extend over a million square kilometers and include such famous sites as Dinosaur National Monument and Como Bluff. The latter site has been described as the single most important discovery in the history of paleontology,[12] and from it the world's museums have been stocked with many of the above-listed dinosaurs. Discoveries here and at other Morrison site continue to be made annually.

Until the 1980s it was widely believed that the Morrison Formation was virtually devoid of tracks despite the wealth of dinosaurs skeletal remains. As with many other formations, we are discovering this is not the case. In fact, there are twenty or more dinosaur tracksites now known, including a particularly large site in the Purgatoire Valley of southeastern Colorado. This site was discovered in the 1930s, forgotten, then reinvestigated in the 1980s; it contains in excess of thirteen hundred mapped footprints, attributable to about one hundred individual animals.[13]

About 40 percent of the trackways represent sauropods, while the remaining 60 percent represent three-toed bipeds, dominantly theropods. The track-bearing beds can be shown to represent a modest-size lake basin subject to periodic phases of drying. It is therefore probable that such sauropod – theropod faunas are representative of Savannah-like, semiarid paleoenvironments with seasonal climatic regimes.[14] As we shall soon see, other evidence suggests that sauropod – theropod communities are often characteristic of similar paleoenvironments at other times in the Age of Dinosaurs.

Now that over twenty footprint sites are known from the Morrison Formation, it can no longer be considered a bone-rich, track-poor formation (category 4 or 5). Instead, one of the world's most famous skeleton-yielding deposits falls into category 3, in which bones and tracks are equally well represented. In fact, there are about one hundred thirty individual trackways known from these twenty sites, a number that compares favorably with the total number of individuals known from complete skeletons.

Early Cretaceous Paleoecology

There are several very important Early Cretaceous tracksite regions. In examining them, we embark on a wide-ranging global tour, taking in the Wealden beds of England, the Enciso beds of Spain, the Sousa beds of Brazil, the Gething Group of Western Canada, the Jindong Formation of Korea, the Glen Rose beds of Texas, and the Dakota Group of Colorado. Without exception, these deposits reveal track assemblages that tell us much about the dinosaur communities. Moreover the evidence they provide gives us information that is largely unavailable from skeletal remains.

The Wealden beds of England

For over a century the Wealden beds of England and similar deposits in other parts of northwestern Europe have yielded numerous tracks attributed to *Iguanodon*. Although few syntheses have been undertaken, the available data suggest a preponderance of *Iguanodon* or *Iguanodon*-like tracks, with only a few theropod footprints.[15] No sauropod tracks are known. The Wealden deposits represent humid, coastal lowland environments with abundant vegetation. The Wealden beds are also well known for yielding the skeletal remains of *Iguanodon*. This makes it a type 3a deposit, where tracks and skeletal remains tell a similar story.

The Enciso beds of Spain

Having commented on the abundance of *Iguanadon* tracks and the rarity of theropod tracks in England, we find in Spain that the situation is reversed in deposits of this age. Theropod tracks are quite abundant and *Iguanadon* tracks are less common. Again, sauropod tracks are virtually unknown. Bones are virtually unknown also, making these beds a classic type 1 deposit, where almost all the evidence of dinosaurs is based on tracks.

The Sousa beds of Brazil

Dinosaur tracks were first reported from the Sousa Formation of Brazil by Giuseppe Leonardi in 1979. Soon after this he realized that he was dealing with a remarkable discovery. The footprints were not confined to one or two layers of strata as was usually thought to be the case in most track-bearing deposits. Instead, he recorded twenty-seven different track-bearing layers in a sequence of strata only about 60 meters thick.[16] This was one of the first well-documented reports of multiple track-bearing layers. As discussed in Chapter 10, we have since learned that many other formations are replete with multiple track-bearing layers.

In order to depict the dinosaur individuals or communities represented by the trackways, Leonardi devised a wonderfully graphic method (see Fig. 8.3). Alongside each track-bearing layer on his geological section of strata he drew in all the dinosaurs that appear to have been represented. Each one was scaled to the appropriate size based on footprint dimensions. This allows us to immediately visualize the community represented by the track census. It turns out that the dinosaur community is dominated by carnivorous theropods with a modest number of ornithopods and a few sauropods. The ornithopod and the sauropod trackmakers do not often appear together at the same levels. The evidence clearly shows that these animals inhabited a lake basin environment for a considerable period of time. Because of the abundance of tracks in this formation and the lack of bones, this is an important type 1 deposit.

Gething Group, Canada

The Canadian Gething Group tracks represent a dinosaur community that probably inhabited an ancient coal swamp, which is an environment of deltas and coastal lowlands quite similar to that represented by the Wealden beds of England. The tracks, notably *Amblydactylus*, an *Iguanodon*-like form,[17] also depict a similar community dominated by large ornithopods with a variety of theropod forms. A single ankylosaur trackway is known and referred to as *Tetrapodosaurus* (Chapter 5). A few true bird tracks have been reported,[18] but sauropod tracks are unknown. Unfortunately the best track-bearing beds in the Gething formation have been flooded by the Peace River Dam. Before this happened, however, over seventeen hundred tracks were documented, and dozens excavated for museums (see Appendix A). The abundance of tracks and lack of skeletal remains makes this an important type 1 deposit.

The Dakota beds of Colorado

Like the Wealden and Gething Group deposits, the Dakota beds of Colorado represent a verdant coastal plain environment where herbivorous ornithopods were abundant. In this deposit two different ornithopod trackmakers appear to be present, including *Caririchnium*, a very common quadrupedal form, and a rarer *Iguanodon*- or *Amblydactylus*-like form.[19] Again, various theropod tracks are known, as well as a few bird tracks, but sauropod tracks are apparently entirely absent from the

SOUSA FORMATION, BRAZIL

Tracksite
Levels

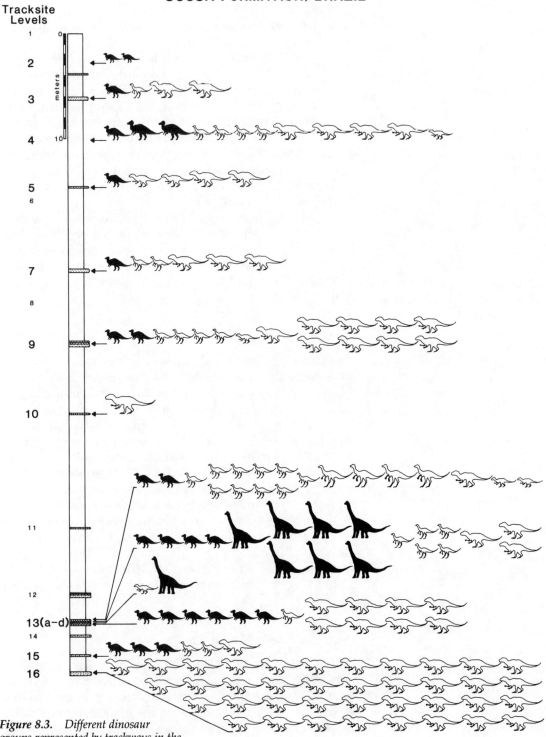

Figure 8.3. *Different dinosaur groups represented by trackways in the Cretaceous Paraiba localities of Brazil. Herbivores shown in black; carnivores, in white. (After Leonardi 1984.)*

CARNIVORES   HERBIVORES

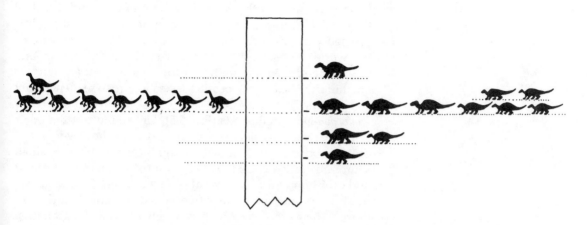

Figure 8.4. *Proportions of different dinosaur groups represented by trackways at different levels in the Dakota Group, Cretaceous of Colorado. Based on Alameda locality.*

deposit. As bones are almost completely unknown from the formation, the Dakota is another classic type 1, track-rich deposit (Fig. 8.4).

The Glen Rose beds of Texas

The famous dinosaur tracks of Texas are about the same age as those from Colorado, but they exhibit a very different paleoecological picture. They occur, not in the sandy and muddy layers typical of swampy, well-vegetated coastal plains, but in limestones and, in some cases, salt-bearing beds, which geologists call a carbonate platform environment.[20] In such environments, like present-day Florida bay or the Persian Gulf, the lagoons became excessively salty from time to time. As water evaporated, the lagoons would shrink and accumulate salt-crusts, gypsum, and other distinctive deposits known as evaporites. As the lagoons gave way to salty, limy mudflats, fresh surfaces were exposed to trackmaking animals. Repeatedly the tracks in these carbonate platform sediments appear to be dominated by sauropods and theropods, with scarce evidence of large ornithopods. In this respect the tracks indicate a dinosaurian community that was quite different from those found in the Wealden, Gething, and Dakota beds. In overall composition the Glen Rose track assemblages are similar to the Late Jurassic Purgatoire footprint assemblage preserved in limy lake beds and the older, Middle Jurassic track-bearing beds of Morocco. The lack of abundant bones in the Glen Rose indicates that it is a type 2 deposit, where tracks are important to our understanding of the paleoecology of this formation.

The Jindong beds of Korea

Imagine we are on the rocky foreshore of the south coast of Korea. All around us are gently dipping layers of strata impressed with distinct dinosaur trackways. This site is one of the world's most prolific dinosaur track-bearing formations (see Appendix A). Preliminary estimates suggest as many as one hundred sixty track-bearing layers in a rock-sequence only 200 meters in thickness.[21] The tracks are mainly those of *Iguanodon*-like ornithopods (Fig. 8.5). Unlike most of the other Lower Cretaceous deposits discussed here, however, these ornithopod tracks occur with a significant number of small sauropod trackways. However, in most cases the sauropod and ornithopod tracks occur at different levels. This suggests these two dinosaur groups frequented the area at different times.* There are also more than twenty levels with bird tracks. A few theropod trackways also occur, but they are notably scarce.

Unlike the vegetation-rich Wealden, Gething, and Dakota deposits, which represent humid coastal plain environments, the Jindong beds represent an ancient lake basin. We can picture a shallow lake whose shores were frequented by dinosaurs and birds. Apart from tracks, the formation yielded few fossils. It is another type 1 deposit, in this case with several hundred individual trackways but no bones.

As we complete our survey of the Lower Cretaceous, some patterns begin to emerge. The herbivorous ornithopods are abundant in some environments, but rare or absent from others, especially those where sauropods are abundant. This could mean different herbivores lived in different environments and overlapped infrequently.

Late Cretaceous Paleoecology

The Winton beds of Australia

We have already briefly examined the Winton beds tracksite at Lark Quarry in Queensland, where a large carnivore allegedly stampeded a mixed flock or herd of small chicken- or turkey-sized ornithopods and theropods. From an ecological point of view, the tracks are a type 2b deposit; the footprints bear no relation to the known skeletal remains of dinosaurs, which consist of only a few fragmentary remains of sauropods. In general, small dinosaur tracks are rare in most Cretaceous deposits, so their occurrence in the Winton beds is a welcome glimpse of small dinosaur activity in an ancient Australian floodplain environment. The two diminutive track-making species may have coexisted in about equal proportions without undue aggression or competition for resources.[22]

*This interpretation has been proposed in a recently published abstract by the author and his colleagues.

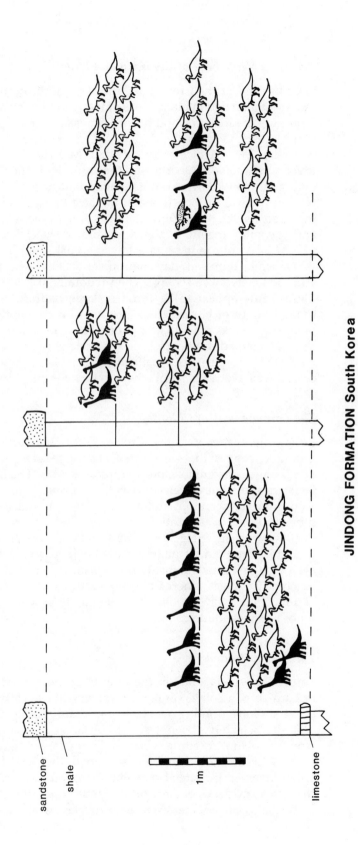

JINDONG FORMATION South Korea

sandstone

shale

1 m

limestone

Figure 8.5. Proportions of different dinosaur groups represented by track-ways in a part of the Jindong Formation, South Korea. All dinosaurs are herbivores, except for a single carnivore (stippled).

The Mesa Verde beds of Colorado and Utah

Near the end of the Cretaceous, the skeletal record of dinosaurs is dominated in many areas of the world by the duckbill
dinosaurs (hadrosaurs) and horned dinosaurs (ceratopsians).
In most cases, the skeletal-rich deposits, for example, the North
American Two Medicine Formation of Montana, the Judith
River Formations of Alberta, and the Lance Formation of Wyoming, are not known to contain abundant tracks.[23] By contrast, further to the south one of the best-known track-bearing deposits, the Mesa Verde deposits of the Colorado, Utah,
and Wyoming region, is almost devoid of skeletal remains.
The Mesa Verde represents a lowland coal swamp setting,
that resembled present-day environments like the Everglades
and the Okefenokee swamp. The predominant tracks in the
Mesa Verde appear to be those of large hadrosaurs.[24] Although bones of horned dinosaurs are also common at this
time, their tracks are rare and poorly known in these deposits. This marked difference from the skeletal record suggests
either that ceratopsians usually avoided these environments
or that we have yet to learn where to look for their tracks.

The Laramie beds of Colorado

Recent discoveries in Latest Cretaceous strata near Golden,
Colorado, suggest that we may have begun to solve the riddle
of missing horned dinosaur (ceratopsian) tracks. The Laramie
beds are sometimes referred to as the *Triceratops* fossil zone.
At this tracksite ceratopsian tracks have been found at several
different levels in association with the tracks of theropod and
duckbill dinosaurs (Fig. 8.6). The formation was previously
thought to contain virtually no fossil evidence of dinosaurs
even though it is rich in plant fossil remains.[25] It is best classed
as a type 2 deposit, where tracks are very important. They tell
of a moderately diverse dinosaur community in which horned
dinosaurs were quite common.

Toro Toro, Bolivia

Latest Cretaceous strata in the vicinity of the remote village
of Toro Toro in Bolivia contain a layer with at least fifty theropod trackways running parallel to eight sauropod trackways.
This site, which has been described by Giuseppe Leonardi as
a marine platform setting (Chapter 7), appears to be another
example of the aforementioned sauropod–theropod fauna.[26]
Other layers at the site also exhibit theropod trackways, but
there is no evidence of large ornithopods at the site.

Although large Late Cretaceous tracksites are less well

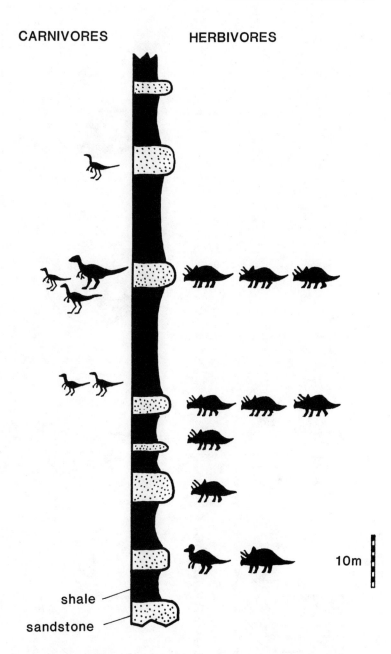

CARNIVORES   HERBIVORES

shale
sandstone

10m

Figure 8.6. *Proportions of different dinosaur groups represented by trackways in a part of the Laramie Formation near Golden, Colorado.*

known than Early Cretaceous examples, there are significant points of comparison. In particular, the Mesa Verde coal swamp tracks tell a story of ornithopod-dominated dinosaur communities resembling those we just examined in the similar plant-rich, coal-bearing coastal plain deposit of the older Wealden, Gething, and Dakota sequences (see Fig. 8.7). This strongly suggests that Cretaceous dinosaur communities dominated by large ornithopods were primarily associated with

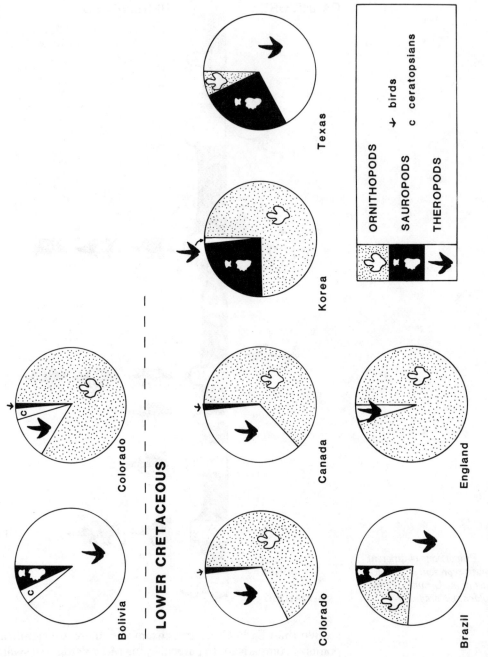

UPPER CRETACEOUS

Bolivia

Colorado

LOWER CRETACEOUS

Korea

Texas

Colorado

Canada

Brazil

England

ORNITHOPODS

SAUROPODS

THEROPODS

↳ birds

c ceratopsians

Figure 8.7. *Proportions of different track types from Cretaceous deposits around the world are used to give an indication of the makeup of dinosaur communities.*

humid, well-vegetated coastal plain environments.[27] By contrast, sauropod-dominated faunas predominated in habitats characterized by alkaline lakes, seasonal aridity, and limy or salty sedimentary deposits.[28] The fact that sauropod track assemblages generally occur in similar settings in the Jurassic supports this generalization. As discussed in the following chapters, these patterns of footprint–sediment relationship are rapidly emerging as significant ecological phenomena in need of analysis and interpretation.

When we study the entire footprint assemblage at a site, we are looking at a bigger problem than a simple statistical tallying of the number of different track types. The tracks represent distinct dinosaur communities – and the distribution of these communities around the world is not random or haphazard. Each community had an ecological relationship to a particular habitat. As an observant paleoecologist once pointed out, "Behind every sedimentary rock there lurks an ecosystem, but what one sees first is an environment of deposition."[29] For our purposes we can paraphrase this, in reverse, and say that behind every census of the dinosaur community is a bigger picture, the ancient environment in which that community existed. We will look at these ancient environments in Chapter 10.

9

Evolution

Feeding birds accompany an iguanodontid dinosaur on the shore of a Cretaceous lake. Artwork by the author.

Those little feet will walk and climb
And run along the road of Time:
They've started out, and where they'll go
'Tis not permitted us to know
– Edgar A. Guest, "The First Step"

Throughout geological time various groups of plants and animals have undergone significant evolutionary change. Plants evolved from algae to roses, animals from worms to dolphins. Dinosaurs were no exception; they existed for almost an entire geological era, approximately 165 million years. During this time they evolved from small birdlike bipeds to giant brontosaurs, hadrosaurs, and ceratopsians. Each new branch of the dinosaur family tree gave rise to innovations. In some cases more efficient teeth, in others robust armor or longer legs, evolved; all of these show up in the fossil record. Other, more subtle changes affected skin color, physiology, metabolism, and various characteristics not so easily determined from fossils, if they can be so determined at all.

The evolutionary changes that affected dinosaur feet and locomotion are fairly obvious. For example, footprints and stride patterns reflect significant changes in the number of toes and changes in gait. These in turn may give us clues about the evolving anatomy of certain dinosaurs in various areas and at particular times.

Because the dinosaur family tree has been established in some detail, tracks are not usually considered particularly important when it comes to determining evolutionary trends that might significantly affect our knowledge of the configurations of the family tree. It would be a bold researcher who substantially challenged the accepted dinosaur family tree on the basis of footprints alone. There is some evidence, however, from footprints that indicates minor modifications in the family tree. More important though, tracks provide other insights into evolution, particularly as it pertains to the evolution of behavior and dinosaur communities.

Before looking at some of the modifications in existing theory that have been suggested, it is worth pointing out that evolution follows a number of paths. One is divergent adaptation. Some families change in appearance through time, particularly if their adaptations diverge significantly. For example among the mammals, bats and elephants are very different despite having a common ancestor. Similarly brontosaurs and birds evolved from a common dinosaur ancestor but progressively adapted to very different lifestyles.

The opposite of this divergent pattern of evolution is known as convergence. When creatures that are only distantly related adopt similar lifestyles or adaptations, their appearance can become increasingly similar. Sharks, dolphins, and ichthyosaurs are very similar in appearance because they have developed streamlined adaptations for efficient high-speed swimming. Yet one is a fish, another a mammal, and the third an extinct reptile.

The phenomenon of convergence affects feet. This is partly because in many animals feet are used for similar purposes – often, although not always, for walking or running, rather than for swimming, digging, or climbing. Most bipedal dinosaurs were three-toed digitigrade toe-walkers, like birds. Similar in many ways to the tracks of carnivorous theropods, the three-toed tracks of ornithopods were made by herbivorous dinosaurs only distantly related to the carnivores.[1] Convergence largely accounts for the similarities that frequently lead to confusion.

Similar problems exist in the differentiation of tracks made by large quadrupedal dinosaurs. These animals had rather large rounded or ovoid, elephantine feet designed to support their large bulk on pillarlike limbs. They were graviportal animals (see Chapter 2). Consequently trackers sometimes confuse the tracks of brontosaurs, ceratopsians, and ankylosaurs.[2]

Despite the problems associated with convergence, tracks may still teach us some lessons about evolution. If we begin at the dawn of the Age of Dinosaurs, we find footprint evidence suggesting an earlier appearance of true dinosaurs than is shown by dinosaurian skeletal remains in the fossil record (Fig. 9.1). Two of the leading proponents of this idea have been Georges Demathieu of Dijon, France, and Hartmut Haubold, from Martin Luther University in Germany. They find distinctly dinosaurian tracks as early as the Middle Triassic, whereas well-authenticated skeletal remains have been recorded, at earliest, in Late Triassic deposits.[3] This is perhaps not surprising; it is improbable that the oldest dinosaur skeleton unearthed would represent the earliest dinosaur.

In general most of the clearly recognizable footprint types appear in the track record when we would expect them to. We rarely find tracks suggesting that particular groups appeared long before their skeletal remains became incorporated into the fossil record. Similarly tracks rarely show that dinosaur groups lingered long after the last evidence of fossilized remains. This inspires confidence in the track record as a reliable reflection of evolutionary trends and ecology.

Let us now look briefly at how we can follow an evolutionary trend through the track record. Dinosaurs represented a

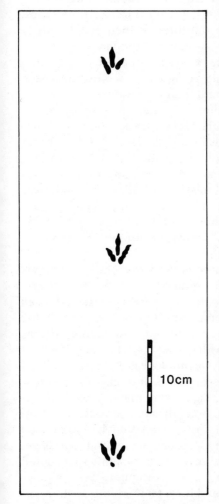

10cm

Figure 9.1. *The oldest reported dinosaur trackway, from the Middle Triassic of France.*

very small component of the animal communities in the latter part of the Triassic. Moreover, based on the track record, relatively small carnivorous theropods were predominant. Following the development of this type of track into Jurassic times, we find an interesting trend toward a number of larger species. In Late Triassic times the largest tracks were typically on the order of 15 to 20 centimeters long, rarely reaching more than 25 centimeters. These are typically referred to as *Grallator* (see Chapter 5). In Early Jurassic times, larger *Eubrontes* tracks appear, some measuring up to 30 to 35 centimeters in length. By the end of Early Jurassic times, large theropod tracks measure up to 45 or 50 centimeters in length (Fig. 9.2).[4]

The track record also tells us about the evolution of locomotion. Dinosaurs evolved from four-footed ancestors to semibipedal and finally to two-footed, birdlike prototypes. Some groups remained bipedal through their history, sticking with the final prototype design. All the birdlike carnivorous dinosaurs, for example, were essentially bipedal. Likewise, the three-toed ornithopods have traditionally been regarded as bipeds throughout their evolution.

In general, footprints confirm our interpretation of carnivorous dinosaurs as bipeds. However, tracks show that supposedly bipedal ornithopods included a large number of representatives that went around on all fours. The earliest presumed ornithopod trackways, *Atreipus*[5] and *Anomoepus* from the Late Triassic and Early Jurassic, indicate animals alternating between quadrupedal and bipedal locomotion (see Chapter 5). As we are now learning, this appears to have been the mode for a number of well-known ornithopod trackmakers from Cretaceous times, too.

Trackways of Early Cretaceous *Iguanodon*-like ornithopods and their Late Cretaceous descendants, the duckbills, clearly indicate that a significant number of species progressed on all fours. [It could be that some ornithopods reverted to the primitive, predinosaurian, four-footed condition, whereas others adopted the new bipedal design (see Fig. 9.3).] In light of the evidence, dinosaur paleontologists have reevaluated their reconstructions and restorations of duck-bills and their relatives, depicting an increasing number in four-footed poses.

In the early Mesozoic (Triassic through Jurassic) most ornithopods were fairly small. Presumably they had to feed on low vegetation. Later in the Mesozoic, many ornithopods became significantly larger, but their predominantly four-footed progression could have encouraged them to remain low-browsers, which in turn would have favored four-footed progression. It has been suggested that the relationship between plant evolution and herbivore feeding styles is reciprocal – that the decline of long-necked, high-browsing sauropods and

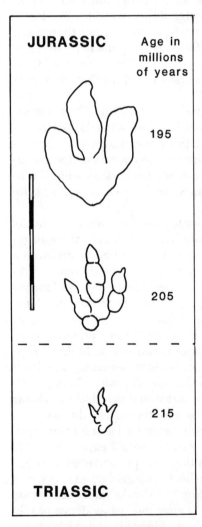

Figure 9.2. *A twenty-million-year trend showing the increase in maximum size of tracks of carnivorous dinosaurs in western North America.*

ORNITHOPODA

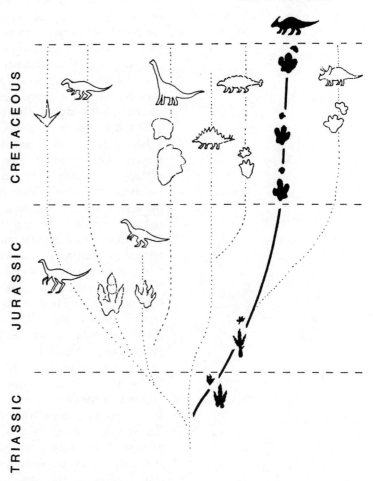

Figure 9.3. Ornithopod tracks reveal a long history of four-footed progression, even though ornithopods are usually thought of as bipedal.

the corresponding rise of low-browsing ornithischians were directly related to the rise of flowering plants, and vice versa. Unlike the earlier seed-bearing plants, such as conifers, flowering plants have the ability to regenerate rapidly even when cropped close to the ground. Some say, therefore, that low-browsing dinosaurs "invented flowers"[6] by causing natural selection to favor fast-growing flowering species, putting excessive pressure on slow-growing seed-bearing varieties. The trackways add to the plausibility of this hypothesis suggesting that a high proportion of four-footed ornithopods were indeed mainly low-browsers, like their other herbivorous ornithischian relatives.

The footprints of sauropods and prosauropods offer little incentive for a reevaluation of traditional views. Some prosauropods may have progressed on two feet and others on four, but, based on early track evidence, all true sauropods were apparently quadrupedal from the time of their first appearance.[7] In fact, trackways of stegosaurs, ankylosaurs, and ceratopsians are very poorly known (Chapter 5) and shed little or no light on development of locomotion. Robert Bakker has suggested that stegosaurs may have been bipedal and advocated their reconstruction in two-footed poses,[8] but as mentioned above, there is absolutely *no* known trackway evidence to help address this hypothesis.

The track record also tells us about the evolution of social behavior. If, as sociobiologists hold, behaviors are passed down genetically, we may be able to use tracks to date the origin of certain genetically determined social traits, such as herding behavior. In a recent paper, Dan Chure, the park paleontologist at Dinosaur National Monument, points out that with the abundance of footprint evidence, we should be able to determine which groups show signs of herding behavior (discussed in Chapter 7) and when such behavior first arose.[9]

We know from evidence at the Mount Tom site, described by John Ostrom, that small theropods were probably gregarious at least by Early Jurassic times. Based on evidence from the Purgatoire site in Colorado, brontosaurs were certainly gregarious by Late Jurassic times; in all probability they were gragarious in Middle or even Early Jurassic times. Certainly the abundance of tracks in the North African strata described by Shinibou Ishigaki suggests large populations, or at least significant congregations, and probable herding activity. The large ornithischians were definitely gregarious from Early Cretaceous times onward. Ornithopod footprints confirm this at many sites from England, Korea, and North America, to name just the best examples. Although we cannot determine anything about the origin of social behavior in stegosaurs, armored dinosaurs, or ceratopsians from an analysis of their tracks (there simply is not enough evidence), ceratopsian tracks occur in significant numbers in the Laramie Formation, as discussed in the previous chapter.

These observations lead to interesting conclusions. First, most major groups of dinosaurs were gregarious, or at least they are represented by species with gregarious habits. Second, based on footprint evidence alone, they appear to have developed their social traits early in their evolutionary history (Fig. 9.4). They were evidently gregarious from their very first appearance, implying gregariousness in their ancestors as well.

Chure suggests we might also use footprints to determine the relative sizes of herds. When we do so, we must remember that the parallel trackways we count at a given site are likely to represent the minimum herd size in that area. For example, we see twenty-three brontosaur trackways at Dav-

Theropoda   Sauropoda   Ornithopoda

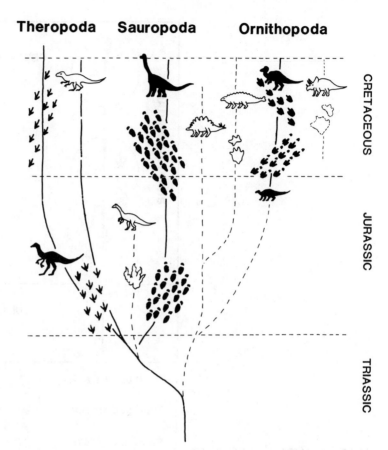

CRETACEOUS

JURASSIC

TRIASSIC

Figure 9.4. Footprints help show when various dinosaur groups developed gregarious (herding) habits. For carnivorous dinosaurs, brontosaurs, and large Cretaceous ornithopods (iguanodontids and duck-bills), tracks indicate that the herding habit became established early in the history of these groups. Horned dinosaurs are also known to have been gregarious, but the evidence is based mainly on bones, not footprints.

enport Ranch, about twenty parallel ornithopod trackways at one level at the Korean site, and over fifty trackways of theropods at Toro Toro in Bolivia, but these almost certainly represent only the minimum size of the groups that passed through these areas. However, we cannot determine how big herds got; from existing trackway evidence, we cannot determine whether some species had larger herds than others or whether the sizes of herds changed over time. Ultimately we will probably need bone and nest-site evidence, as well as footprint evidence, to attempt such determinations.

Finally, the track record yields information about community evolution. In Chapter 8 we showed that the general composition of dinosaur communities can be inferred from the track evidence and that distinct community types are associated with particular environments. For example, we demonstrated beyond reasonable doubt that the Early Cretaceous Wealden, Gething, and Dakota deposits are dominated by *Iguanodon*-like ornithopod tracks and that similar duckbill footprint assemblages occur in comparable Late Cretaceous

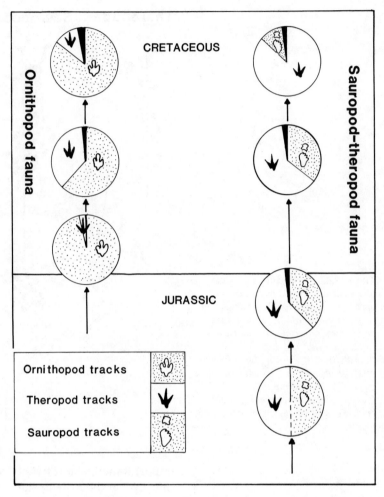

Figure 9.5. *Tracks show proportions of different types of dinosaurs through time. Note the persistence of two distinct communities.*

deposits representing similar ancient environments. This strongly suggests that Cretaceous ornithopod-dominated communities evolved an ecological preference for well-vegetated, humid, coastal plain environments. By contrast the sauropod-dominated track assemblages are usually characterized by a high proportion of theropod tracks but a distinct lack of ornithopod tracks. Moreover, they usually occur in environments that represent limy and salty lakes and coastal lagoons.

Let us attempt to show this evolution graphically. Taking the diagrams from Chapter 8 that show proportions of different dinosaur types, we can then plot them in a time sequence, as shown in Figure 9.5. Of course, we want to distinguish the ornithopod-dominated communities from the sauropod–theropod communities because they occurred in different environments. We can trace the sauropod–theropod communities from Early and Middle Jurassic through Late Cretaceous, and the ornithopod-dominated communities through most of the Cretaceous.

The pattern makes evolutionary sense. The sauropods evolved in the Jurassic and fed on primitive vegetation, such as seed-bearing plants and ferns.[10] Evidence suggests that they mainly inhabited semiarid or seasonal climates, rather than humid, wetland settings. They were probably preyed and scavenged upon by carnivores that followed sauropod migrations.

In Cretaceous times, when the large ornithopod-dominated communities became established, there was no reason to expect the sauropods to suddenly abandon their long-established habits and preferences for certain foods and environments; they apparently stuck to what they preferred: drier inland environments.[11] That they did not try to compete with the large ornithopods that inhabited the humid coastal lowlands makes ecological sense because it allows resources, particularly food, to be shared or partitioned.

Large ornithopods may have been better adapted for browsing on low vegetation. As time passed, such vegetation would have included an increasing number of the new, rapidly growing, flowering plants. Although ornithopods must initially have eaten primitive plants (like ferns and the seed-bearing varieties), by Middle and Late Cretaceous times they are commonly associated with deposits that contain abundant evidence of flowering plants; presumably they took advantage of this abundant new food source.[12] Because sauropod tracks are absent or rare in most of these deposits, it seems unlikely that they availed themselves of this new resource to any great extent. Indeed, the fact that they failed to do so and stuck to their old dietary preferences may have contributed to their decline. They may have become increasingly restricted to environments dominated by the "old" vegetation.

Dinosaurs did not evolve in isolation. They shared their Mesozoic world with other groups of vertebrates, including amphibians, other reptiles, and birds. As mentioned earlier, most of these other groups have a poor track record, either because they were small creatures or less abundant at this time, or both. However, recent discoveries show that one group, the birds, has a significant track record. Birds are also thought to have a close evolutionary relationship to dinosaurs.

Dinosaurs and birds

Paleontologists have long known about a relationship between dinosaurs and birds. Names like *Ornithischian*, meaning "bird-hipped," or *ornithopod*, meaning "bird-foot," demonstrate this point. However, until recently the notion that birds descended directly from dinosaurs has not been widely accepted. Could *Tyrannosaurus rex* really be the great-grandfather of a hummingbird? In a roundabout way, the answer is probably yes! Work by John Ostrom in the early 1970s and by Jacques Gauthier in the 1980s has shown that birds apparently descended directly from theropod dinosaurs. As one author recently put it, "[B]irds are dinosaurs."[13] This apparent relationship leads to some amusement if we substitute "dinosaur" for "bird" in everyday speech – "A dinosaur in the hand is worth two in the bush"!

A discussion of dinosaur – bird relationships is outside the scope of a book on tracks. However, despite controversy, there is evidence that birds are related to dinosaurs; paleontologists just cannot agree on exactly what the relationship is. This need not deter us. We can proceed with an examination of the important insights that tracks reveal about bird evolution during the Age of Dinosaurs.

Assuming that pigeons and chickens may really be theropods, we need to consider when birds became birdlike and distinct from their dinosaurian relatives. Our modern view of birds is of feathered friends, in most cases capable of flight. They are more birdlike if they have these modern traits. Until recently the traditional view has been that *Archaeopteryx*, from the Late Jurassic, was the first bird, a conclusion supported by the fact that this very dinosaurian creature had feathers. Of course, the oldest known bird is not necessarily as old as the earliest bird. In paleontology we are always finding ancestors of creatures that were previously thought to be the oldest examples of a particular group. The same has recently happened with the study of bird fossils. We know of an Early Cretaceous bird that is more birdlike than *Archaeopteryx*, as well as a protobird called *Protoavis* from older Late Triassic deposits.[14]

What do tracks tell us? The traditional view was that tracks are rare and the oldest bird tracks date from the Cretaceous Period.[15] Recent discoveries suggest, however, that tracks are quite common and that the track record may date back to the Early Jurassic.

In Chapter 8 we saw that bird tracks are quite common in Early Cretaceous deposits. Sites in Colorado, Canada, and South Korea are among a growing number of important localities with abundant bird tracks (Fig. 9.6). All these tracks appear to have been made by water birds that left footprints resembling those of modern ploverlike waders. This is not surprising because these birds frequent shorelines where they make lots of tracks, often while feeding; perching, tree, or

TRACKS OF CRETACEOUS BIRDS (Theropods)

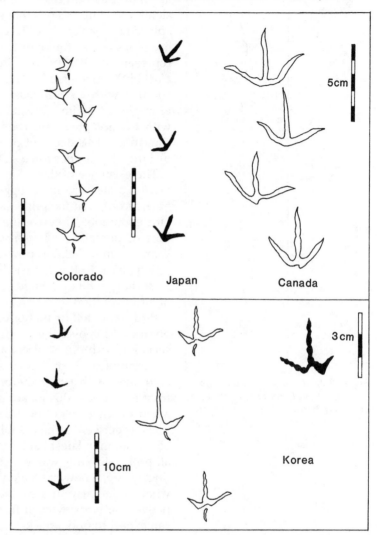

Figure 9.6. Various Early Cretaceous bird tracks from North America and eastern Asia indicate that a diverse group of waterbirds was then in existence.

rock-dwelling birds are less likely to leave abundant tracks because they are not in the habit of alighting on wet ground.

If abundant waders and waterfowl had evolved by Early Cretaceous times, where did they all come from? Did they all descend from *Archaeopteryx*? The answer is probably no. A significant phase of earliest, or pre-Cretaceous, bird evolution was probably required to give rise to such a diversity of waterfowl. The lack of Early Cretaceous skeletal remains of water birds, and of birds in general, also suggests that the track record has a lot to offer our understanding of bird evolution.

There are at least two, possibly three, known examples of

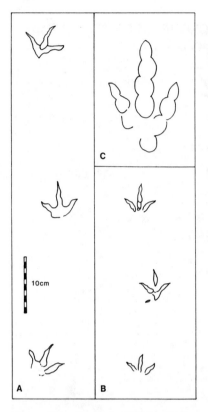

Figure 9.7. *Example of bird or bird-like tracks from Early Jurassic strata, with a* Grallator *track (c) for comparison. Other very bird-like tracks of this age are known from Southern Africa.*

Early Jurassic trackways that are very birdlike (Fig. 9.7). Unlike most dinosaur tracks, these footprints are very small, with slender toe impressions that show that the digits were widely splayed out, a very birdlike characteristic. The two illustrated examples come from North America and North Africa; they represent small animals, with feet no bigger than a modern blackbird's or pigeon's. The North American example is associated with a desert oasis setting, and the North African example with a limy coastal plain.[16] Comparing these tracks with the accepted Cretaceous bird footprints, we find clear similarities. This does not prove the Jurassic tracks were made by birds, but it does force us to consider this possibility.

The existence of these Early Jurassic bird or birdlike trackways is quite exciting. It suggests, as does the Cretaceous bird-track record, that footprints can add to our knowledge of early bird evolution. They are the oldest birdlike tracks so far reported, predating *Archaeopteryx* by as much as fifty million years. Even if *Protoavis* proves to be a true bird, pushing bird origins back to the Triassic, the Early Jurassic tracks will be important in filling a huge gap at a crucial time in the history of bird evolution.

Bird tracks tell us that birds evolved during the Age of Dinosaurs. Moreover, birds appear to have evolved in close association with dinosaur communities. Both the Lower Jurassic examples cited herein are of birdlike tracks that are associated with larger theropod footprints. If birds evolved from theropods, this makes sense; at some point early in the Mesozoic there were no true birds, only theropods. Theropods underwent a remarkable evolutionary radiation and diversification in latest Late Triassic and Early Jurassic times. In all probability this was when birds began to evolve as a distinct group, a new branch on the family tree. This is probably what these Early Jurassic footprints are showing us. (Alternatively, if birds evolved from predinosaurian ancestors, we can expect to find a track record beginning early in the Mesozoic.)

By Early Cretaceous times we have several examples of sites where bird tracks occur in association with abundant dinosaur tracks. In the Colorado, Canada, and Korea examples cited above, the water bird tracks are primarily associated with ornithopod-dominated communities that inhabited lake and river shorelines (Fig. 9.8). Although we cannot read too much into the coexistence of birds and dinosaurs, we should be aware that birds were a part of many Cretaceous dinosaur communities and that their tracks from this period are by no means rare. Conventional wisdom holds that there was a Late Cretaceous radiation of water birds, about seventy million years ago. This conclusion is based largely on the availability of

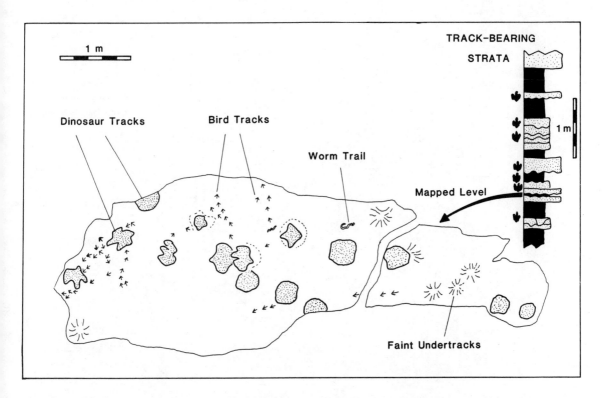

Figure 9.8. *Bird and dinosaur tracks together on a single slab. (See also Fig. 4.6.)*

skeletal remains at the time and their scarcity in older, Early Cretaceous deposits. Dozens of bird track reports now prove a widespread radiation of waterbirds at least thirty million years earlier (by one hundred million years ago). Even the skeptic has to accept that tracks shed much light on the evolution of bird communities at this time. The aforementioned, Lower Jurassic, bird-like tracks suggest another potentially surprising insight into bird evolution. If the trackmakers were indeed birds, then we must consider the possibility that they originated at least 50 million years before Archaeopterix.*

As we shall see in the next chapter, to date we have nothing like a complete picture of dinosaur communities. Our conclusions need testing against more data from tracks and other sources. The evolution of communities is not controlled only by the evolution of various groups or by their proclivities for particular food sources and particular climates. Other factors, such as the ancient geography, were important in controlling animal distribution and evolution. Because these factors relate to the ancient environment as much as they do to ecology or evolution, we shall deal with them under that heading.

*At the time of writing the authors and his colleagues have a manuscript in review on "The Track Record of Mesozoic Birds"

10

Dinosaur tracks and ancient environments

Large dinosaurs like Camarasaurus *were responsible for trampling lake shore sediments around Dinosaur Lake, southeastern Colorado. Artwork by Doug Henderson.*

Dinosaurian populations in general were strongly influenced by the local and regional landscape.
— Giuseppe Leonardi, *Dinosaur Tracks and Traces,*
(1989)

During the age of dinosaurs it was not just the biological world of plants and animals that was so vastly different from what we know today; the physical world, the environment, was different. The continents were not in the positions they are in today; there were inland seas in areas that are now dry land, and there was dry land in areas that are now submerged. Every river and lake had specifically Mesozoic configurations. The positions of mountain ranges, shorelines and climatic zones were different from those at present; moreover, they changed constantly through the successive epochs of geologic time. It is against this backdrop of unfamiliar landscapes that we must view the dinosaurian world.

Although these ancient landscapes have disappeared, traces of them remain in sedimentary accumulations, testifying to the former existence of lakes, rivers, beaches, and other ancient habitats. Subtle clues in the composition and chemistry of the sediments and the diversity of fossil remains help us reconstruct the ancient environments and their climatic regimes. Also remaining, of course, are the tracks and trackways of the inhabitants of these lost worlds.

Until now we have adopted a biological approach to tracking. We have mainly assumed that tracks are useful in providing information about the trackmaker, such as size, speed, direction of travel, variety of species, ecology, and evolution. In this chapter we turn to a new area of scientific investigation, one not traditionally considered part of track studies: paleoenvironmental research, or investigation into the ancient environment.

A few years ago, when the paleoenvironmental importance of tracks was becoming more widely recognized, I suggested that footprints could be useful in understanding environments inhabited by dinosaur communities, patterns or cycles of sediment accumulation, and particular ancient environments and shoreline trends.[1] Tracks can also help us estimate water depths, the water content of sediments, and current directions, and determine ancient slopes or gradients in the landscape. Furthermore, tracking reveals how footprints and trampling disturbs ancient soils, plants, and animals.

In just a few years our knowledge of the context and preservation of tracks in the rock record has greatly improved.

Because of the scope of the subject, our discussion will run to three chapters. We begin with a broad overview of ancient environments and then proceed to examine particular environments in detail.

Mesozoic world geography and animal communities

In the past two chapters evidence was presented for various dinosaur communities apparently having strong preferences for particular habitats or environments. We briefly mentioned rift valleys, deserts, lake basins, humid swamps, and limy coastal plains. We also saw that evidence of ecosystems is found within the track-bearing sedimentary strata; one needs time and experience to dig and tease it out.

When we speak of finds in Spain or England or Colorado, it is understood that the modern setting does not resemble the prehistoric environment. As shown in Figure 10.1, the Mesozoic world was quite different from the one we know today. It is generally accepted today that the continental masses are forever on the move, riding on the huge tectonic plates that make up the earth's crust; in each geological period the configuration of continents and oceans is different.[2] For example, throughout the Age of Dinosaurs, a large low-lying ocean called Tethys separated Africa, the Middle East, and India from Asia.[3] Moreover, early in the Mesozoic there was not yet an Atlantic Ocean separating the New World (the Americas) from the Old (Europe, Africa, and Asia). By the end of the Age of Dinosaurs, however, the Old and New Worlds had begun drifting apart.

The close fit of land masses seen in the Early Jurassic map of Figure 10.1 had been a feature of the face of our planet from the Late Paleozoic (immediately preceding the Age of Dinosaurs). This supercontinent, known as Pangea II, permitted free interchange of animals and plants, with the establishment of widespread, similar communities. It is generally accepted that life forms were cosmopolitan, or pandemic.[4] In other words, they were controlled to some degree by the configuration of the supercontinent. Other environmental factors also came into play, however. Doglike Lower Jurassic tracks known as *Brasilichnium* are found in desert deposits in North America (Fig. 10.2), South America (Brazil), and southern Africa. Although widespread in one sense, these trackmakers were also endemic to desert environments. This particular example is important because it appears that mammal-like reptile tracks are often found in desert deposits, not just in the Late Jurassic, but back into Permian times, long before the Age of Dinosaurs. For example, the track-rich Permian deposits of the Grand Canyon area are almost identical in many respects to those of the Early Jurassic deposits that accumu-

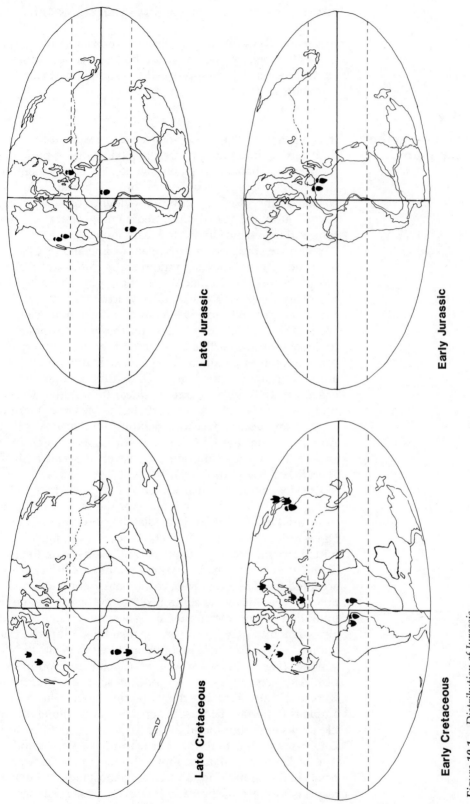

Late Cretaceous

Late Jurassic

Early Cretaceous

Early Jurassic

Figure 10.1. Distribution of Jurassic and Cretaceous brontosaur footprint localities, showing their predominance at low latitudes. Ornithopod tracks occur at higher latitudes.

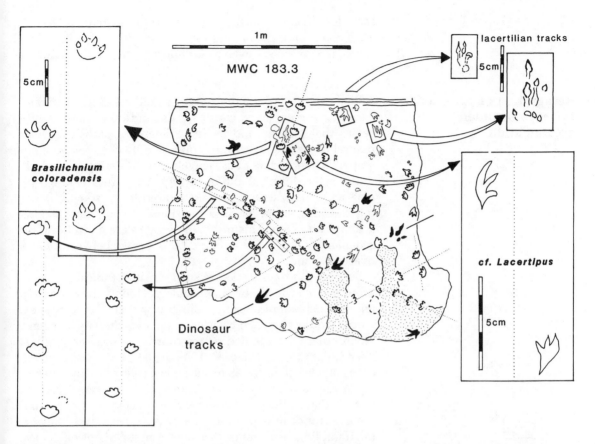

Figure 10.2. *A slab of Lower Jurassic desert sandstone with many mammal-like* Brasilichnium *tracks. MWC = Museum of Western Colorado specimen.*

lated almost one hundred million years later. If it were not for the presence of dinosaur tracks in the Jurassic deposits, even an experienced tracker would be hard pressed to distinguish the deposits based on footprints alone. Once certain groups of mammal-like trackmakers became established in desert environments, it seems, they remained there for a long time.[5] They never expanded into the great many ecological niches occupied by the successful dinosaurs. The desert was probably a haven or preferred habitat for them.

By the Early Jurassic, however, the record shows, dinosaurs had made inroads into the desert environments. And at about this time, too, the mammal-like tracks and skeletal remains disappear. Evidently, the mammal-like forms were eventually squeezed out of their refuges, perhaps by the dinosaurs or perhaps by other pressures.

In Middle and Late Jurassic times Pangea II was breaking up. Animal communities were becoming separated and endemic; there were more divergent evolutionary forms. For example, from their first appearance in Early Jurassic North Africa right through to Cretaceous times, brontosaur tracksites are associated with limy substrates in tropical and subtropical latitudes. In contrast, the ornithopod-dominated Cretaceous assemblages are mainly associated with middle to high latitudes, in some cases even as far north as the Arctic circle.[6]

Recognition of patterns or cycles of sediment accumulation

Sedimentary layers accumulate as a result of the action of river currents, waves, tides, storm winds, and other natural processes that pick up and move sedimentary particles. These natural processes are in turn controlled by climate and landscape. When sediment is actively accumulating, especially underwater during flooding or high tide, trackmaking animals are usually unable to make footprints. As soon as flooding subsides, however, and the tide goes out, trackmaking activity can begin. This means that tracks are an indicator of sedimentary cycles and in particular of breaks between phases of accumulation.

Because dinosaur tracks are such sensitive indicators of wet ground, water level, or water table position,[7] they can help tell us much about ancient sedimentary cycles. A simple example would be a lake basin like the ancient Dinosaur Lake environment represented by the Morrison Formation deposits of Colorado (see Chapter 8). As the lake level rose and fell, dinosaurs left their tracks in successive layers. The number of tracks found in any given layer and area reflects at least two factors, that is, the number of dinosaurs active in the area and the amount of time the substrate was exposed and available for trampling. Assuming that over long spans of geological time dinosaur populations displayed something like equilibrium in numbers, we can attribute much of the variation in trampling intensity to the duration of time between the accumulation of layers. This gives us an important general principle: Track accumulation in a given area is largely controlled by the length of time between periods of sediment accumulation. Several lines of evidence support this principle.

At the Dinosaur Lake site, tracks occur at four successive layers representing shallow or low-water phases in the lake's history (Fig. 10.3). Two of the layers are fine limy mudstone that accumulated slowly; the others are limy sandstone deposits that accumulated rapidly, probably as the result of storms or floods. It appears to be more than mere coincidence that the slowly accumulated layers are very heavily trampled and the rapidly deposited beds have a much lower density of tracks.

This principle is well known in geology. For example, the degree of disturbance by burrowing, or bioturbation, of marine sediments by invertebrates is considered in part a function of time.[8] Similarly the intensity of trampling, or dinoturbation,[9] gives us a possible measure of time. Tracks and trampling occur at the tops of beds and indicate a break in the

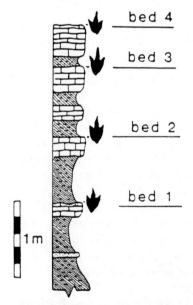

Figure 10.3. *The regular pattern of footprint-bearing layers in a sedimentary sequence indicates cycles of sedimentary deposition and climatic change. Morrison Formation, Dinosaur Lake locality, Colorado.*

sequence of deposition of a sedimentary succession. In effect, during these hiatuses it is only the tracks that accumulate.

Sedimentary accumulations are cyclical in many cases. When a lake level drops, tracks are made in the moist lake basin sediments exposed. As the lake level rises again, the track-bearing layers are blanketed and preserved by new deposits of lake sediment. The Sousa Formation of Brazil has yielded twenty-seven track-bearing layers in a sequence of 60 meters, and the Jindong Formation of Korea has yielded about one hundred sixty track-bearing layers in a succession only 200 meters thick. Similar track-rich successions have been identified in the Triassic–Jurassic rift lake successions of the North American eastern seaboard (Newark Series). At most of these sites track-bearing beds occur at intervals on the order of only 1 or 2 to perhaps 10 meters. Track-bearing layers seem to be a persistent, rather than an occasional, feature in many geological successions.

Although it is easy to recognize cyclical deposition patterns when layers occur with regular frequency, it is not always obvious what the cycles mean. In the study of the Triassic–Jurassic lake successions of the Newark Series the sequence of sedimentary cycles has been linked to long-term climate fluctuations controlled by complex variations in the earth's rotation and orbit. These include cycles on the order of 25,000, 44,000, 100,000, 133,000, and 400,000 years,[10] which are in close agreement with present-day cycles of roughly 21,000 (precession of the equinoxes), 41,000 (obliquity cycle), and 95,000, 123,000 and 413,000 years (eccentricity cycles).

Even when dinosaur footprint beds occur with less regularity, they reflect cyclical sedimentary deposit. There are extensive track-bearing beds at single horizons, for example, that resulted from changes in sea level. Such changes are directly or indirectly linked to climate regimes. These megatracksites are associated with major hiatuses or interruptions in sediment accumulation (see Chapter 12).

Recognition of particular ancient environments

Recognizable tracks can be made only on soft or receptive substrates that yield and deform with the trackmaker's footfalls. Such substrates are usually wet, occurring at or near the watertable adjacent to rivers, lakes, or oceans or over broad lowland areas, such as swamps, wetlands, or floodplains. In the latter settings there may not be such a clear distinction between water bodies and areas of land, however, the principle of track formation is the same: The receptive substrates are at or near the watertable.

In rare cases tracks can be preserved on dry substrates. Tracks are encountered in dry, sand dune environments. Even

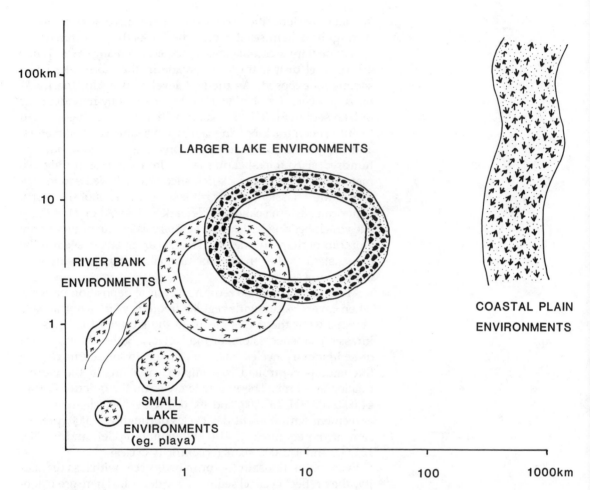

Figure 10.4. The size and shape of dinosaur track-bearing layers is controlled by the ancient environment. In theory and in practice these layers vary from very localized concentrations (for example, rivers and playa lakes) to larger circular or subcircular configurations (large lakes) to extensive elongate ribbon zones (marine shorelines, coastal wetlands, and swamps).

though sand dunes are easily blown away, which would reduce preservation potential, tracks can also be preserved. As we saw in Chapter 3, footprints deforming a buried layer may survive while the surface expression of the same tracks is obliterated. This principle permits wet substrate undertracks to be preserved while the surface tracks are washed away – or blown away, in the case of sand dunes. (Such partial obliteration of tracks is frequently observed in the swash zone of beaches.)

Given the broad spectrum of environments in which tracks can be formed, it may appear hard to recognize specific paleoenvironments. However, there are a number of clues, including the characteristics of the rocks themselves, that help us distinguish different track-bearing paleoenvironments. In a study of tracks made by large modern mammals in East Africa, Leo Laporte and Kay Behrensmayer found tracks most abundantly in the "relatively narrow zone of moist, vegetation-free sediment" along the shores of Lake Turkana and other lakes.[11] In deposits representing deserts or semiarid paleoenvironments, where the water table occurs only locally at the surface, footprints have a very limited distribution, reflecting the lack of moist, receptive substrates. For example, many fossil playa lakes reveal track-bearing layers that cover only an acre or two. In contrast, megatracksites, where the water table is at or near the track-receptive surface over wide areas of lowland coastal plain, may contain tracks over hundreds or even thousands of square kilometers.

The intermediate cases are too numerous to discuss in detail. They include zones that are narrow but laterally extensive, associated with marine shorelines, and others that show circular or subcircular configurations, associated with lake basins of variable size (Fig. 10.4).

Traditionally paleontologists have used diagnostic fossils, such as those of clams, or snails, to reconstruct ancient environments. Certain species are associated with beaches or intertidal zones, whereas others occur subtidally or in very deep water; some are freshwater forms, others marine. In an analysis of the distribution of tracks in relation to classic ancient environment reconstructions, Jeffrey Pittman demonstrated that most tracks are found in the intertidal zone.[12] Because tracks usually occur in very narrow shoreline zones, they are similarly useful for diagnostic purposes. This is especially important when other fossil evidence is uninformative about an environment.

Pittman's observation offers us the opportunity to determine that deposits were *not* laid down in deep water. Where present, tracks may settle lengthy geological debates over the significance of body fossils and other evidence. Moreover, much may be at stake in correctly identifying an ancient environment. For example, in the work done by Pittman, the ancient environment reconstructions have been used extensively in oil exploration, and before tracks were found some geologists had mistaken shoreline sediments for deep-water deposits.

Recognition of shoreline trends

It is a common observation that animals, including human beings, frequently walk parallel to shorelines. The shore of a lake, the ocean strand line, and the riverbank are the natural demarcation lines separating land and water. As shown in Chapter 7, a strong preferred-orientation trend in a set of trackways can be interpreted as animals following a physically controlled pathway, such as a shoreline, (2) a herd moving in one direction, or (3) a combination of both. In recent years, as more trackway data have come to light, a number of sites yielding strong preferred-orientation trends have been documented. At many of these it has been possible to use supplemental evidence of ripple marks, track depth, sediment type, and so on, to demonstrate that trackways run parallel to ancient shoelines. Thus tracks have paleogeographic significance, because in many ancient deposits the original position of the strandline may not otherwise be clear.[13]

Probably the best examples documented to date are found at the Purgatoire River site in Colorado, where trackway trends and track depths provide complementary evidence of ancient shoreline configurations. Trackways deepen from south to north, so it is easy to infer a body of water or wet substrate to the north (Fig. 10.5). When the depths of sauropod tracks are measured over the whole area, this pattern is generally confirmed, and it is possible to plot an inferred west–east shoreline for the western and central part of the site. Toward the east the shoreline swings northward, suggesting an overall pattern of a northwest facing bay or embayment. Several lines of evidence support this lake embayment interpretation. First, the trackway orientations consistently predominate in an east–west direction along the western segment of the shore, while to the east there is additional evidence, from an underlying track-bearing bed, that the shoreline ran north–south. Second, there is a small channel flowing into the V made by the contours of the embayment, the exact location where a drainage would be expected (Fig. 10.5).

The Purgatoire site is exceptional and has been studied in detail; however, the occurrence of shore-parallel trackways at fossil footprint sites is not unusual. We do not yet know the proportion of sites at which strong shore-parallel trackways can be recognized, but we know that they occur along ancient river channel systems and marine shorelines, as well as beside lakes. Two good examples of trackway trends following ancient river channel orientations have been documented from the Late Triassic Dockum and Chinle Formations of the western United States (Fig. 10.6). The former is particularly obvious because the predominant trackway trends can be seen to follow the general north–south orientation of an ancient winding river system. Throughout much of the tracked area a 220- to 230-million-year-old exhumed channel is as clearly visible as if it had been recently incised. Where the trackways are not following the banks, they can be seen to deepen suddenly as they cross the wet slopes and floor of the channel. Similar river-parallel trackway trends have been recorded in Triassic deposits from France and elsewhere.

In addition to the lake and river examples, we can round out the picture with ocean shoreline examples. The Cretaceous Dakota deposits of Colorado represent an ancient ocean shoreline with a north–south orientation. In several instances

Figure 10.5. Top. *The distribution of tracks of various depths (in centimeters) can be shown on a contour map and used to indicate shoreline position (based on the Purgatoire site). Note the small drainage flowing into the embayment in the expected position. Bottom. Track-bearing sediment, accumulated alongside "Dinosaur Lake," an ancient body of fresh water in Late Jurassic times. The small drainage channel can be seen on the left side of the figure.*

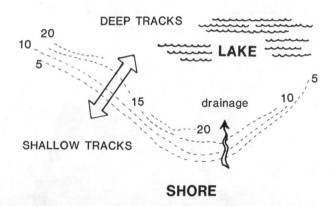

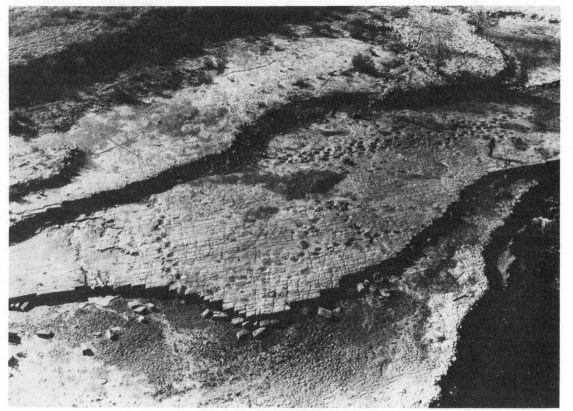

the tracks appear to be following this trend. The same pattern has been reported from trackway sites in the Cretaceous Glen Rose Formation, where animals were apparently walking along marine and lagoon shorelines.

Estimation of water depth

A number of studies have claimed to recognize tracks made underwater by partially bouyant dinosaurs. Dinosaurs were shown almost out of their depth, in 2 to 4 meters of water,

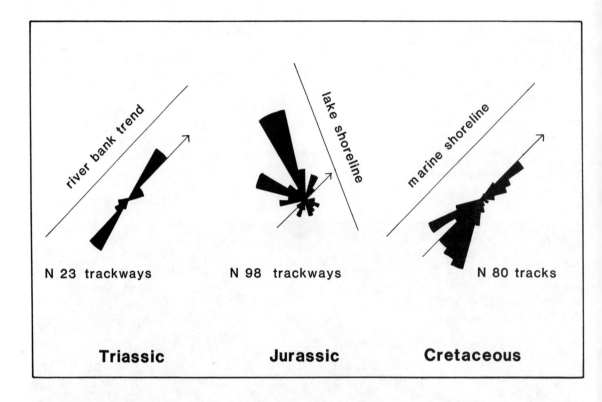

Figure 10.6. Trackway orientations as indicators of shoreline trends. Examples from Triassic, Jurassic, and Cretaceous of Colorado and Utah show trackway trends following various known river, lake, and coastline trends.

touching the bottom only with their toe tips. If these reconstructions of dinosaur paddling activity are correct, then the tracks are a useful and accurate means of estimating water depth at the time the so-called swim tracks were made. If the tracks are incorrectly interpreted, however, the estimates of water depth are almost certainly wrong, and any interpretation of the ancient depositional environment based on them may be consequently incorrect. The reader will see that we have been returned to the swimming dinosaur debate. We aired it first in our discussions of individual behavior (chapter 6) and then in chapters 8 and 9 we considered the ecological and evolutionary implications of global sauropod track evidence. Here, the third time around, it is worth considering the paleoenvironmental implications.

Roland T. Bird was the first to suggest that a water depth of about 3 meters helped buoy up a Cretaceous brontosaur in the ancient Gulf of Mexico. This interpretation is strongly disputed because the possibility of underprints was not considered. Similar interpretations of African sauropod tracks were also challenged for the same reasons. Regardless of which interpretation is correct, it is clear that our paleoenvironmental reconstruction will be very different depending on which hypothesis we favor. If we follow Bird, we will have to recon-

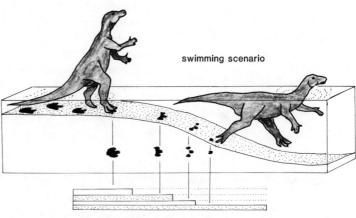

swimming scenario

true tracks undertracks scenario

Figure 10.7. *Interpreting and misinterpreting tracks as indicators of water depth.*

struct the Cretaceous shoreline well inland from where the purported swimmers were active. If we prefer the underprint interpretation, the shoreline would be further oceanward in the opposite direction, maybe by as much as 40 to 50 kilometers! (This is because coastal plains are almost flat, sloping at less than 10 centimeters per kilometer.[14])

Although the sauropods in question were probably not swimming in several meters of sea water, this does not mean that all such interpretations are incorrect. In 1980 Walter Coombs claimed to have found tracks of a Jurassic theropod tiptoeing through about two meters of water in a freshwater, rift valley lake environment. If Coombs is correct, he has used tracks to provide an accurate estimate of lake water depth.

Reconstructing water depths is a prime objective in many types of ancient environment research. Tracks may be helpful, but they can also be misleading if not interpreted with caution. Polish researchers Gerard Gierlinski and Agata Potemska inferred the activity of a partially buoyant dinosaur based on Lower Jurassic tracks from Poland.[15] They depicted a dinosaur heading out to sea (Fig. 10.7). Such behavior seems unlikely, however, especially for a terrestrial herbivore. More to the point, the interpretation is not supported by the footprint evidence, which consists of a mixture of complete and incomplete tracks. These authors admitted that there were no trackways showing a transition from complete to incomplete tracks, that is, showing a dinosaur progressing into deeper water. Despite the lack of evidence, they purported that the complete and incomplete tracks could be rearranged to reveal the desired pattern. They made their evidence fit a behavioral interpretation rather than an ancient environment reconstruction.

It appears likely that dinosaur tracks provide reliable evidence of water depth only in rare cases. The example cited by Coombs may be evidence of swim traces, but those cited by Bird, Ishigaki, and Gierlinski and Potemska do not stand up to scientific scrutiny. Other dubious dinosaur swim tracks are discussed at the end of this chapter.[16] Most of the more convincing examples of swim tracks are associated with aquatic animals like turtles, crocodiles, and fish, but in these cases water depths are not estimated because the animals could have been swimming either at the surface or along the bottom.

Estimation of water content of sediment

Dolf Seilacher, one of the world's leading trace fossil experts once wrote that a footprint is "an experiment in soil mechanics."[17] This principle is intuitively obvious, and we can confirm it every time we observe fresh tracks in the mud after a rainstorm or flood. The presence of crisp, deep tracks reflects a high water content in the sediments, whereas the presence of shallow, almost imperceptible tracks reflects a firm, unyielding substrate with little or no moisture.

Although tracks have not been systematically used to determine the water content of ancient sediments, they have considerable potential for this type of investigation. Since it is possible to estimate the weights of dinosaurs with reasonable accuracy,[18] it follows that we should also be able to use tracks to determine the mechanical properties of the sediment at the time the tracks were made. Such determinations are likely to be of considerable value to sedimentologists and geologists attempting to reconstruct the ancient environment. Many longstanding debates about particular sedimentary deposits revolve around whether climates were humid or arid or whether particular deposits accumulated under water. For instance, in the Cretaceous Dakota Group of Colorado the dinosaur track-bearing layers have been described as well-drained levee deposits,[19] yet the tracks, through several successive layers, are consistently deep. Such evidence suggests a wetter or poorly drained substrate similar to that found in many coal swamp deposits.

Debates over sediment saturation levels have also figured prominently in discussions of how tracks are preserved in desert sand dune environments. Because dry sand is loose, several researchers claim that it is hard to form crisp tracks unless the substrate has been moistened by rain or dew. Some authors have even claimed that tracks in sand dunes must have been made underwater.[20] Clearly, interpretations of a flooded landscape are at variance with our usual concept of a desert as an arid place.

Because tracks are so frequently associated with wet substrates, they are sensitive water table indicators. They are usually formed only when the water table (or water level) is near the substrate. This can be during or after a rainstorm or flood when the substrate is submerged or wet. Tracks are more likely to be made in the waning flood stage or after the flood. As waters are rising, and during peak flood stage, there is too much erosion, and sediment is actually being moved along and over the substrate. Animals too may be less active, taking refuge away from the floods. But as the flood subsides and sediment comes to rest, fresh new sand and mud deposits are laid down as track-receptive substrates, and animals resume normal activity. We can make the analogy of an unexposed film or blank sheet of paper ready to be exposed or marked. Geologists call rising sea or water levels a transgression, and falling or receding water levels a regression.

Since ichnologists have demonstrated that tracks are frequently made in regressive situations, it is possible to predict the occurrence of tracks with greater consistency and to use tracks to identify regressive deposits. For example, when the tide begins to ebb, tracks can be made in the intertidal zone. Those made early in the ebb phase, near the high water mark, will last for almost the entire twelve-hour tidal cycle, whereas those made during the transgressive flood phase cannot last more than six hours before being inundated. On average, tracks made during the ebb phase last two to three times as long as those made during the flood phase. In a lake, where waters may recede very slowly during a prolonged dry season and rise rapidly during and following wet season rains, the proportion of tracks made during the regressive phase may exceed those made in the transgressive phase by an order of magnitude or more. If the regressive phase tracks are preserved as underprints or by being covered with transgressive phase sediment, the proportion of regressive phase tracks preserved will also be very high.

A steep sandy beach is not a good model for track preservation over long-term rising and falling of sea level because tracks get washed out so rapidly with each tidal cycle; a low-energy, low-gradient mudflat provides a better model for much larger coastal plain systems. Such vast areas are subjected to long-term sea level transgressions and regressions, like prolonged flood and ebb phases of the tidal cycle. From what we have seen to date in the geologic record, regressive track-bearing sediments may be more abundant than those of the transgressive phase. This generalization needs to be qualified and tested, however. Sea level drops, or regressions, are eventually followed by rises, or transgressions, and tracks can be made either as the water level drops or later as it begins to rise again. In either event, a sea level drop precedes track-making activity.

In conclusion, the slow lowering of the water table, exposing fresh wet sediment (as on mudflats) across receding lake basins or expanding coastal plains, provides an ideal situation for tracks to be made in abundance and subsequently preserved. By contrast, rising tides, filling lake basins, and receding shorelines do not usually produce sediments that contain tracks. In some cases they may deposit sediments that cover the track-bearing layers of the earlier regressive phase. In others they may erode the footprint-bearing layers.

Determination of current directions

Considering the problems associated with recognizing tracks made underwater, it comes as no great surprise to find that footprints indicative of current activity are equally problematic and often hard to interpret. The problem is essentially similar to that encountered in trying to understand swim tracks. If an animal is buoyed in shallow water barely touching the bottom with its toe tips, it can also be moved or drifted along by currents or wave action. If the animal drags or scrapes its feet along the bottom, the orientation of the toe scrape marks should reveal the current direction.

This principle has been fairly convincingly demonstrated for invertebrate animals like trilobites drifting in weak currents. The principle should apply to larger animals, like dinosaurs of various sizes, which could drift under the influence of more powerful currents. For example, large birds like seagulls drift along in very shallow water with their feet in contact with the substrate. This behavior is a deliberate feeding strategy, designed to disturb invertebrates in the substrate. The gulls shovel through the shallow sea bottom, leaving distinctive traces.[21]

I know of only one example of such a trace in Mesozoic deposits (Fig. 10.8); it comes from a Lower Jurassic dinosaur tracksite in Utah. In this case the trace can only be described as birdlike. There is no firmly accepted evidence of birds at this time (see Chapter 9). The trace shows what appears to be zig-zag forward-and-backward motion perpendicular to the wave ripple marks as if an animal were riding the waves. The trace then moves off the north, shoveling the sediment with its two feet. Finally, it turns through 180 degrees, then drifts off leaving a broad arc-shaped trace. This interpretation is only preliminary. However, many features of this unusual trace suggest a floating, foraging animal with two active legs, if not a bird then perhaps a small dinosaur engaged in birdlike behavior.

To date only a few examples of drifting animals have been reported from Mesozoic deposits, and only some of these have been attributed to dinosaurs. In 1984 Boyd and Loope reeval-

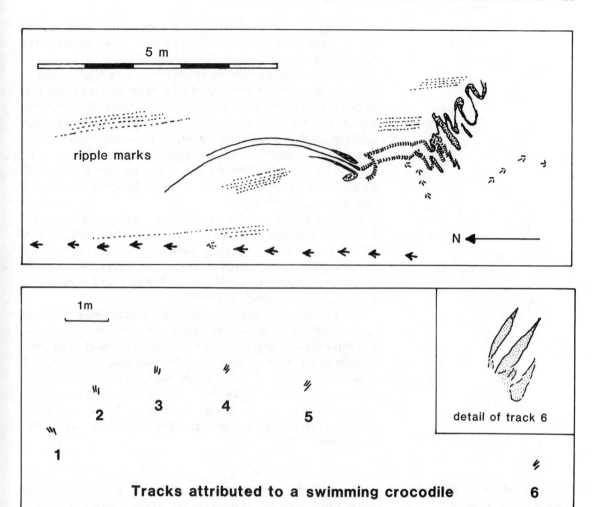

Figure 10.8. *An elongate trace that resembles bird feeding marks, made in shallow water, Jurassic, Utah.*

Figure 10.9. *Toe drag traces suggest swimming vertebrates in submerged environments. This example is a controversial trackway from Cretaceous beds in New Mexico.*

uated elongate drag marks from Triassic sediments in Wyoming that had previously been interpreted as driftwood marks and concluded that they were made by swimming vertebrates probably influenced by a current.[22] Problematic elongate toe traces in the Triassic Chinle Formation of Utah have been interpreted in the same way, although it appears impossible to prove the hypothesis one way or another.

In the Lower Cretaceous of both New Mexico and Kansas there are a number of perplexing tracks of long-toed animals. These have been variously attributed to pterosaurs taking off,[23] to swimming crocodiles (Fig. 10.9), and to swimming dinosaurs. In the case of the New Mexico tracks a crocodile drifting with a current seems likely. The Kansas tracks, whether belonging to partially buoyant ornithischian dinosaurs or crocodiles, are aligned in one consistent orientation, possibly indicating that the trackmakers were influenced by a current that caused them to drift and scrape their toes parallel to the current's direction.

The final example comes from the Lower Cretaceous of Canada, where Phil Currie reported a hadrosaur trackway that appears to represent an animal walking through about two meters of water.[24] "At one point the midline of the trackway shifts more than a meter to the right," he writes, "and several steps later, it shifts to the left again." Such shifts could be caused by currents flowing broadside against the animal, though one would expect the shifts to be consistent in the downstream direction.

Currie also described a now famous example of a set of at least five hadrosaur trackways that swerve in unison (see Chapter 7). These swerving trackways were made as the trackmakers crossed what was apparently a very wet and soft, probably submerged, substrate, and they could suggest the presence of a broadside current.

In summary, it appears that buoyant animals may drift with currents, leaving tell-tale drag marks. At present the evidence for dinosaurs doing this is sketchy and inconclusive even though better evidence exists for other animals, such as aquatic crocodiles. It should be noted, however, that current direction indicators like ripple marks are quite common in close association with dinosaur tracks and can at least tell us which way currents flowed at about the time dinosaurs were active in the area.

Determination of ancient slopes or gradients

Despite the fact that strata often have a layer-cake appearance, indicating that they were originally deposited as horizontal sheets, some sedimentary rock units were deposited on inclined surfaces. Examples include beaches, slopes of river deltas, and desert sand dunes. Here, the dip or inclination of the strata can be quite high, up to 25 or 30 degrees. Even apparently flat-lying deposits may have very low, imperceptible gradients on the order of 1 degree or less.

Fossil animal tracks including those of dinosaurs are not found on the slopes, or foresets, of deltas, because such deposits accumulate underwater. The foreset (i.e., front set) is the inclined slope of a delta or sand dune as it builds forward. On sand dunes the foreset is the steep slip face, or avalanche face, on the lee, or downwind, side. However, tracks are known to occur in various beach deposits and on the slipfaces of desert dunes. The inclination of dune foresets can be as high as 30 degrees. This means that if an animal is walking upright, it will leave an asymmetrical footprint characterized by a bulge of displaced sediment on the downslope side. This

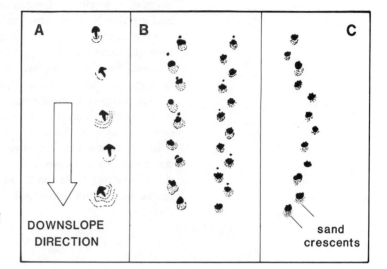

Figure 10.10. *Sand crescents or bulges on the downslope side of footprints allow geologists to reconstruct the slope of ancient sand dunes. A: dinosaur tracks from California; B and C:* Brasilichnium *from Brazil and Utah, respectively. All from Jurassic sand dune deposits.*

bulge has sometimes been referred to as a sand crescent, or impact rim. Because gravity dictates that the sand crescent should normally be on the downslope side, it is possible to use tracks to determine the original direction of inclination of the dune face. In one study a slope of 25 degrees was estimated.[25]

Tracks with sand crescents behind them indicate an animal progressing upslope and are probably the most common orientation encountered in sand dune environments (Fig. 10.10). Tracks with sand crescents in front of them indicated animals moving downslope and are rarer. This is probably in part due to the fact that animals traveling downslope are generally moving faster and in doing so can displace more sand and cause miniavalanches that obscure their tracks. Examples of animals moving horizontally along the contours of a dune are also known and can be recognized by the presence of sand crescents along the sides of the footprints.[26]

Instances of dinosaurs walking on slopes in other environments are not documented with any certainty. Even though beach and lakeshore paleoenvironments are identified with confidence, significant slopes or gradients of more than a few degrees have not been measured accurately or documented in detail. On occasion authors have suggested that tracks with deep toe impressions and no heel marks indicate animals progressing up a slope. This is certainly a possible interpretation, but there are two other equally probable explanations. First, like other animals, dinosaurs would dig their toes in more deeply when progressing rapidly at a trot or a run; such gaits should be recognizable from long step and stride lengths.

Second, toe-only footprints are characteristic of underprints and are much more common than previously supposed.

Finally, it is worth mentioning that extensive tracksites or megatracksites represent low-lying coastal plain environments where the gradient or paleoslope is imperceptible or negligible. These sites are discussed in detail in Chapter 12. The occurrence of tracks over such wide areas is in itself evidence of the existence of a very low gradient where the land surface and water table could coincide.

Before turning to these megatracksites, we need to examine one more phenomenon – trampling. For over 150 million years dinosaurs left their mark on the Mesozoic soils, flora, and fauna that went underfoot. This is the subject of our next chapter.

11

Trampled underfoot

A herd of brontosaurs reconstructed from trackway evidence at Davenport Ranch, Cretaceous of Texas. Artwork by Edward Von Mueller.

Keep your eyes open for tracks and signs of . . . ploughed up pond bottoms.
– Clyde Ormond, *Complete Book of Outdoor Lore* (1964)

Let us now look at what happened when dinosaurs trampled and disturbed the earth beneath their feet. As can be imagined, a large herd of large animals can have considerable impact, not only plowing through and churning up the soil, but also disturbing or destroying plants and even the animals that live in the substrate.

Taphonomy is the study of what happens to animal and plant remains between death and final burial. It is a relatively new branch of paleontology that has revealed much interesting information on how bones and other remains get transported, trampled, and distributed, after the death of animals. This information helps us understand the ancient environments in which the animals lived and died and shows what circumstances lead to ideal fossil preservation (minimal disturbance before burial) and poor preservation (considerable disturbance before burial).

The same taphonomic principles that affect bones or body fossils can apply to tracks, albeit with some subtle but significant differences. Crisp, clear dinosaur tracks can be preserved in pristine condition and appear as fresh looking as if they were made yesterday. Alternatively, they may be affected by wind, water, or the influence of other animals (trampling) and deteriorate to a state of poor preservation. However, unlike bones, tracks may be of poor quality at the time they are made and thus be hard to distinguish from originally well-preserved tracks that have deteriorated significantly. For this reason it is important to carefully study the context of such tracks (their relationship to the enclosing sedimentary layers) to determine if they have always been of inferior quality or if they deteriorated after they were made. Such study should help reveal a distinction between the properties of the substrate at the time the tracks were made and the changes that affected the substrate after the tracks were formed.

Such distinctions might sound subtle and academic on paper, but they are fundamental. A skilled tracker of modern animals needs an understanding of changes affecting footprints to determine how long ago particular tracks were made. An experienced tracker can tell exactly when an animal passed by and even what has happened since. Moreover, trackers can usually identify specific animals as the trackmakers, even if the evidence is indistinct and poorly preserved.

Another factor in footprint taphonomy is the difference between true tracks and undertracks, or track infillings. Undertracks may resemble true tracks that have been modified by water or other disturbance ("taphonomic alteration," in the jargon of the new science).

One way to begin making these various distinctions is by identifying true tracks, such as those with skin impressions, that have undergone little or no taphonomic alteration. Current evidence suggests that these are sometimes preserved when sand fills crisp tracks made in firm, cohesive clayey substrates. The track is effectively acting as a sediment trap. This point is neatly demonstrated by the occurrence of footprints cast in sandstone that originate from strata in which no sandstone beds exist. The only explanation appears to be that sand-laden currents washed across the track-bearing clayey substrate and into another area; only in the deep indentations formed by the tracks could sand accumulate. The footprints formed pockets where energy conditions were low, and these acted as sediment traps (Fig. 11.1).

In the Late Jurassic Purbeck beds of England dinosaur tracks acted as traps for the remains of small animals.[1] The concentration of fossils in the footprints led to the preservation of material that might not otherwise have been fossilized. Sediment was washed into the tracks along with bones. This helped bury potential fossils in a localized protected environment that did not exist outside the footprints.

Studies indicate that modern tracks also act as traps for small bones, dung, and seeds and often make good germination sites. For example, well-worn bear trails on the North American tundra often grow rich clumps of grass where seeds have collected in the footprints.[2] Similarly, after the Mt. Saint Helens eruption of 1980, fresh tracks acted as traps for seeds and moisture, thus providing germination sites and helping plants recolonize the devastated landscape. Because fossil tracks are essentially the same shape as modern tracks, they also act as traps for water and biotic remains, as well as for sediment. It is not unusual to find dinosaur tracks that are filled with water, soil, and sprouting plants. On coastlines dinosaur tracks even form small seaweed- and invertebrate-filled tide pools along rocky foreshores. Studies of modern footprints have also revealed that they are sometimes used as nest sites by ground nesting birds. On an otherwise flat marsh or mud flat a track is an ideal protected indentation in which to gather nest material and raise a clutch.

Roland Bird reported that Cretaceous dinosaur tracks act as

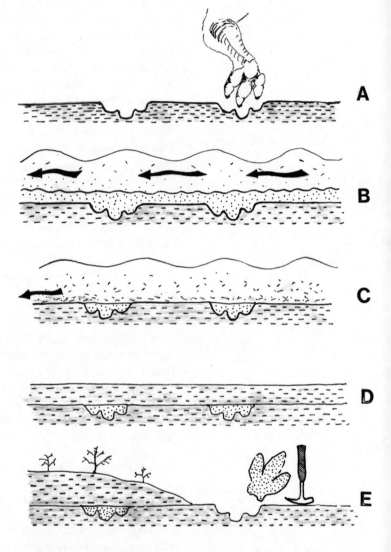

Figure 11.1. *Tracks act as sediment traps, in some cases preserving sediment not seen elsewhere in the local sequence of strata.*

traps for catfish in the Paluxy River. Thus the bones found in fossil tracks could, in some cases, conceivably represent remains of animals that were trapped while alive. Presumably such a fate, which could apply to stranded fish, would not befall four-legged creatures capable of climbing out of even the deepest tracks. However, small vertebrates may be trampled by large vertebrates and end up at the bottom of large footprints! Around certain east African lakes, for example, large populations of frogs exist in areas regularly trampled by hippopotamus and other large mammals. It is not uncommon to find dead frogs that were killed by trampling.

Thus, deep dinosaur tracks created pockets conducive to improved preservation potential, not just for the track itself, which soon filled in, but for the infilling material, whether it was sediment alone or sediment enclosing dead or trampled animal and plant remains. When such tracks are reexhumed, they may again act as sediment traps and influence the local ecology.

Dinosaurs also improved preservation in the fossil record by trampling and burying plant and animal remains that would not otherwise be preserved. One of the more complex geological processes is the hardening, or lithification, of sedimentary deposits to form sedimentary rock. This involves processes known as diagenesis, cementation, and compaction, which begin when the sediment is first deposited and continue for thousands, even millions, of years afterward. By the time a sedimentary deposit is buried a few hundred feet below overlying layers that have accumulated subsequently, it is thoroughly compacted under the pressure of a thick layer of overburden and well on its way to being fully hardened, or lithified. These overburden forces are known as geostatic and hydrostatic pressure. Essentially they are proportional to the weight of the overlying sediments and water, respectively, at any given point. Because we know the weight (density) of these materials, the overburden pressures can easily be calculated. Since we can calculate the pressure exerted by dinosaur feet if we know the foot size and approximate weight of the animal, we can also calculate the amount of compaction caused by various dinosaurs and estimate its effect relative to normal sediment accumulation processes. We can calculate a new type of force – *dinostatic* pressure!

Recently, the pressures exerted by the large brontosaurs *Diplodocus*, *Apatosauras* (= *Brontosaurus*), and *Brachiosaurus* have been calculated. *Diplodocus* weighed anywhere from 10 to 20 tons; *Apatosaurus* and *Brachiosaurus* reached 35 to 55 metric tons.[3] Large as they were, their back feet were little more than half a square meter in area. This means they exerted as much as 30 to 40 metric tons pressure per square meter on the substrate when simply standing. When walking or running, their feet would have pounded the substrate more forcibly, increasing these pressures significantly, easily doubling them.

We can now consider how these dinostatic pressures compare with the natural overburden pressures, the geostatic pressure caused by the normal processes of sediment accumulation. Wet sediment is about twice as heavy as water. This means that a cubic meter (a little more than a cubic yard) weighs about 2 tons. If we imagine a square meter on the earth's surface, the pressure at 1 meter's depth is 2 tons per square meter; at 2 meters it is 4 tons per square meter, and so on. We feel about half this pressure when we dive to corresponding depths below the water surface. In order to reach pressures

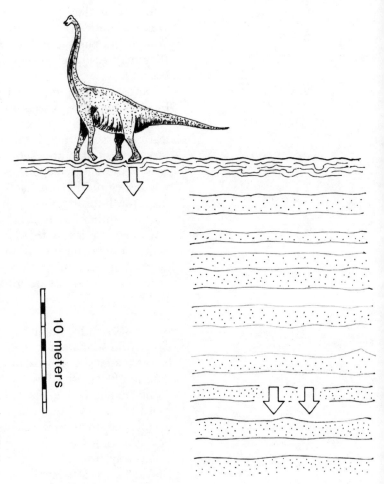

Figure 11.2. *The pressures produced by large dinosaur footfalls are equivalent to being buried by 15 to 20 meters of overburden.*

of 30 to 40 tons per square meter, we would have to be buried 15 to 20 meters below the ground surface (Fig. 11.2). At such depths sediments are subject to considerable compaction; the rock-forming processes of cementation and diagenesis are usually well underway. When dinosaurs trampled a substrate, they exerted comparable pressure right at the surface, and so left a very distinctive mark on certain layers of strata.

Although we usually think of dinosaur tracks as indentations, they quite often appear as raised areas because the sediment where the dinosaurs trod has been compacted and become more resistant to erosion than the sediment in the surrounding areas. Subtle color changes often accompany these footprint compaction features. Such cases confirm the tremendous compaction effect caused by dinosaurs. The principle is the same as in the compaction of snow. Often in winter we see icy footprints preserved in areas where the surrounding snow has blown away or melted. Similarly, along modern

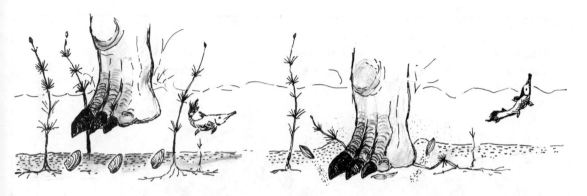

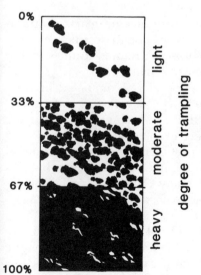

Figure 11.3. Brontosaur trampling killed several dozen clams in the bed of a 150-million-year-old lake, Morrison Formation of Colorado. Trackway shown below; reconstruction of trampling event depicted above.

dead clams

1m

brontosaur tracks

DINOTURBATION INDEX

0%

33%

67%

100%

light

moderate

heavy

degree of trampling

Figure 11.4. The amount of substrate trampled by dinosaurs can be measured using the dinoturbation index.

lakeshores tracks and undertracks may stand out as compacted raised areas around which uncompacted sands and silts have been blown away.

Since dinosaurs had such a significant impact on the substrates they trampled, we would expect them also to affect plants and animals associated with these substrates. In 1862 it was reported that an *Iguanodon* had disturbed *Cyrena* clam shells in the lower Cretaceous Wealden beds of England. Another example of clam carnage was reported from the Jurassic Morrison Formation of Colorado.[4] In this instance a brontosaur had trampled and killed at least two dozen clams in the space of a few strides (Fig. 11.3). Modern studies reveal that livestock can trample and injure or kill freshwater clams. When these poor creatures valiantly regenerate their broken shells, they produce healed but mishappen forms that have been referred to as "molluscan monstrosities."[5]

Because plants were ubiquitous (and immobile), they were also subject to considerable trampling by dinosaurs. Like clams, they certainly could not get out of the way. At the same Colorado sites where the clams met their death, there are abundant plant remains trampled into the limy muds of the lakeshore environment. It is well known that rapid burial is one of the more important factors contributing to good fossil preservation. As dinosaurs trampled plant stems, roots, fronds, and foliage, they forced them from the surface to levels sev-

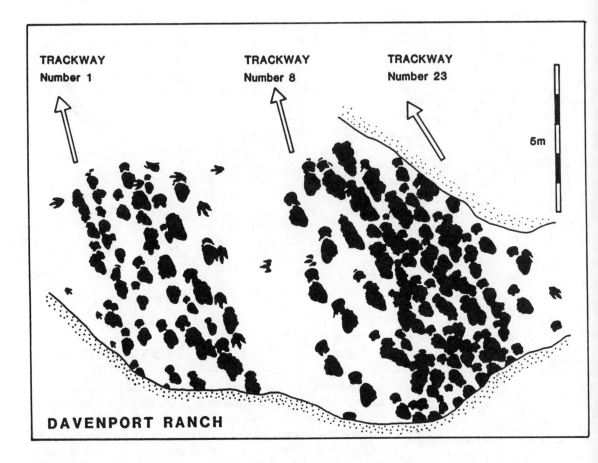

TRACKWAY
Number 1

TRACKWAY
Number 8

TRACKWAY
Number 23

5m

DAVENPORT RANCH

Figure 11.5. *A herd of Cretaceous brontosaurs trampled up to 30 percent of the substrate as it passed. Example from Davenport Ranch, Texas.*

eral centimeters down. They were making major contributions to the rapid burial process. Although destructive to the wildlife around them, dinosaurs contributed to the creation of fossils with their heavy-footed activities.

As we now know, many deposits accumulated in well-vegetated environments, such as coal swamps and various delta plain–coastal plain settings. In such settings dinosaurs may have trampled thousands, even millions, of plants every day. It used to be considered a student blunder to regard coal as the remains of vegetation trampled by dinosaurs, but this is in fact what happened in many Cretaceous coal swamps. We now know that Mesozoic coalbeds contain extensive evidence of trampled vegetation.

The trampling process should not be underestimated. Although it has not been studied in any detail, it is becoming clear that many Mesozoic substrates or soils were extensively affected by dinosaurs. As mentioned in the previous chapter, this process is called dinoturbation, that is, the bioturbation attributable to dinosaurs. It has been recognized widely in deposits throughout the Age of Dinosaurs and can be mea-

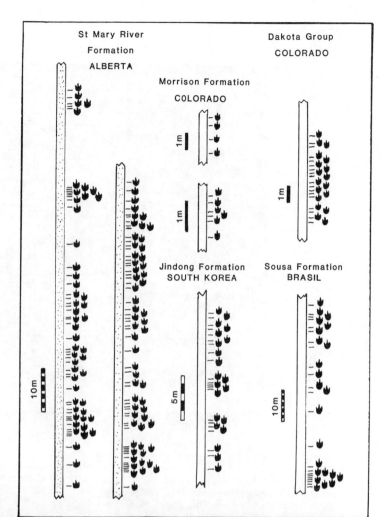

Figure 11.6. Some sequences of strata are replete with dinosaur track-bearing layers. Much of the Mesozoic substrate was trampled underfoot at one time or another.

Figure 11.7. Exceptionally deep brontosaur tracks, Purgatoire site, Colorado. Single track (left), trackway (right).

Figure 11.8. *Dinoturbation in Jurassic and Cretaceous sediments. Purgatoire site, Colorado (above), Jindong site, Korea (below).*

sured to give an indication of the extent of dinosaurian impact.[6] As shown in Figure 11.4, we can recognize where various proportions of the substrate have been affected and call substrates lightly, moderately, or heavily trampled. We have adopted the term *dinoturbation* to propose a dinoturbation index.[7]

For example, a single herd will trample a given percentage of the substrate when crossing a particular area. In the measured example from Davenport Ranch (see Chapter 7), twenty-three brontosaurs trampled 20 to 30 percent of the substrate while crossing an area 15 meters wide (Fig. 11.5).[8] Larger herds and repeated activity clearly result in a greater degree of substrate impact. Charles Darwin observed that the soil surface of the entire English countryside had been bioturbated by passing through the stomachs of lowly earthworms; just so, the mighty dinosaur left no soil unturned or unchurned in many ancient Mesozoic settings. Dale Russell even used the term "ploughed" to describe the extensive alteration of soils caused by dinosaurs and other large vertebrates.[9]

This extensive impact of dinosaur trampling is now beginning to be recognized in many deposits from the Mesozoic Era. Many formations from around the world reveal multiple track-bearing and trampled layers, showing that dinosaurs plowed or disturbed substrates repeatedly in certain environments. Some of the most spectacular examples known to date come from Jurassic and Cretaceous deposits in North America, Korea, and Brazil (Figs. 11.6 – 11.8). In these examples dozens, even hundreds, of track-bearing layers are stacked one on top of another. The rock strata are replete with trampled, tracked, and disturbed layers.

Geologists have already identified localized and extensive dinoturbation in several dozen ancient deposits. Because of their large size and herding activities, many late Mesozoic dinosaurs, particularly the brontosaurs and large ornithopods, may have affected the substrate underfoot to a greater degree than at any other time in the history of vertebrates.[10] As we shall see in the next chapter, this is not mere speculation. Certain sedimentary deposits bear the unmistakable traces of widespread trampling.* Layers of strata are churned and disturbed over large areas. The mighty dinosaurs made big and lasting impressions.

*The author has a paper in press (in Spanish) on this subject. The title translates as "Dinoturbation and the Phenomenon of Vertebrate Trampling in Ancient Environments."

12

Megatracksites:
a new era in tracking

*Three iguanodontids walk along a
Cretaceous beach, Denver, Colorado.
Based on trackway evidence from the
Dinosaur Freeway. Artwork by John
Sibbick.*

*That turnpike Earth! – that common highway all over dented
with the marks of . . . heels and hoofs.*
   – Herman Melville, Moby Dick *(1851)*

Dinosaurs trampled large areas of the Mesozoic landscape,
but how extensive was the impression they left behind? As
explained in Chapters 10 and 11, this depends on the abundance of dinosaurs and the configuration of wet substrates in
ancient environments. Sometimes conditions conspire to produce truly gigantic tracksites.

Megatracksites are footprint-bearing layers of strata that
cover large geographic areas on the order of hundreds, even
thousands, of square kilometers. They have been recognized
only in recent years, and their significance is not yet fully
understood.[1] By definition they are very extensive and different from more localized tracksites, which are known to extend for only a few tens or hundreds of square meters. (In
fact, some local tracksites may prove to represent only small
parts of much larger track-bearing layers that are seen only
sporadically at the surface. As more and more tracksites are
discovered, it becomes increasingly evident that many tracksites are only parts of a bigger picture. The cases in which the
limits of tracksites are completely or even partly known are
the exceptions, not the rule.)

It is helpful at the outset to remind ourselves once more
that the majority of tracks are made on wet substrates where
the water table is at or near the surface. Such substrates may
be localized around desert lakes and certain water courses or
extensive, for example, as in lowland coastal plain environments.[2] The track-bearing environment gives us a preliminary clue to the probable extent of track-bearing layers (see
Chapter 10, Fig. 10.3).

As discussed above, the distribution of tracks is localized
around small lakes and water courses, leading to the development of small tracksites. Larger lake basins may lead to the
development of much more extensive track-bearing layers.
However, it appears to be the low-lying, low-gradient coastal
plain environments that are conducive to the formation of the
most regionally extensive track-bearing layers, or megatracksites. Such environments were like the present day Gulf of
Mexico coastal plain and extended, in some cases, for hundreds
of thousands of square miles. This preliminary inference is
based in part on the observation that all three known megatracksites represent such ancient coastal plain environments.

The oldest megatracksite currently known is situated in

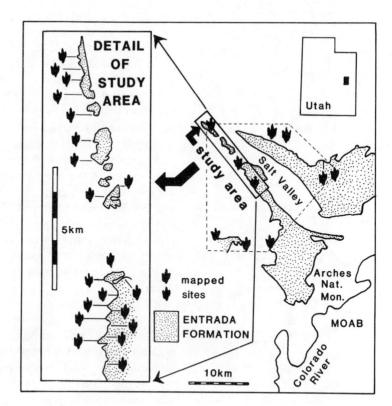

Figure 12.1. *The Middle Jurassic Entrada Formation megatracksite of eastern Utah. Concentration of individual mapped tracksites in western part of study area is shown on the left. Dashed line shows known distribution of track-bearing layer.*

Eastern Utah and is Late Middle to Late Jurassic about 160 to 150 million years in age. The tracks occur at the top of the Middle Jurassic Entrada Formation. The main body of this formation represents an ancient sand sea of the type associated with the Sahara desert today. During Lower and Middle Jurassic times, much of the southwestern United States was dominated by these sand sea environments, but in Late Jurassic times the environment began to change. The overlying Summerville and Morrison formations are known to be floodplain deposits with various river and lake sediments.

The area that is now eastern Utah was periodically encroached on from the west by shallow seas that reworked or redistributed the sediments of the desert margins into lowlying coastal plain deposits, variously interpreted as coastal mud and sand flats, lagoons, and lakes. A distinct lack of fossil remains and accumulations of gypsum salts suggest that conditions were not favorable for the development of diverse plant and animal communities.

It is in just such a setting, where the Sahara-like desert sediments merge with coastal plain deposits, that we encounter extensive footprint evidence. A road sign in northeastern Utah identifying the distinctive Entrada Sandstone has a subtitle,

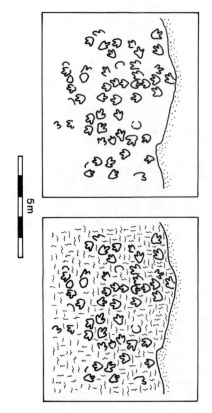

Figure 12.2A. *A dense concentration of tracks from one small area within the Entrada Formation megatracksite. Top diagram shows only clear tracks. Lower diagram shows how areas between tracks are disturbed by trampling.*

"No Sign of Life Here," yet not far away to the south the formation yields millions of dinosaur tracks.

Tracks at the top of the Entrada have been known from the vicinity of Moab, Utah, since about 1940. In recent years local residents had reported a number of sites that had never been documented in the literature. Upon further investigation it became clear that these sites were all in the same layer at the very top of the Entrada where it is in contact with the overlying Summerville Formation.[3]

Beginning in the summer of 1987 a systematic survey of these outcrops was undertaken, and maps of areas with well-exposed tracks were compiled. After a while it became clear that tracks could be found almost anywhere that the top of the formation was exposed. The top of the Entrada is exposed almost continuously in one area, in a belt about 15 kilometers long and a couple of kilometers wide. The outcrop can be walked out over this area for 30 square kilometers (Fig. 12.1). After several days of reconnaissance in this area dozens of mappable sites had been recorded on aerial photographs. It was then safe to report that the tracks form a more or less continuous carpet or zone at this level – a dinosaur stomping ground of enormous proportions. To further test this conclusion and help define the extent of this zone, tracks in another area some 15 to 20 kilometers northeast of this 30-square-kilometer patch were sought, and four new sites were found in the space of an afternoon. When we combined these areas, the extent of the track-bearing layer was established at about a minimum of 300 square kilometers.[4]

Based on detailed mapping at selected sites, it is possible to estimate at least one footprint per square meter (Fig. 12.2 A, B). In places the layer is heavily dinoturbated, allowing for estimates of at least ten tracks per square meter. If the density of tracks averages between one and ten per square meter over the whole area, then we can estimate between three hundred million and three billion tracks. Such an abundance of footprints clearly qualifies as a "sign of life."

One of the first questions people ask is, why such a rich concentration of tracks? Looked at from a nongeological perspective, the high density of tracks might be taken as evidence of very large dinosaur populations, possibly tens or hundreds of thousands, even millions, of animals frequenting the area for a short period of time. But in the same way that many animals can make a large number of tracks in a short period of time, so a smaller number of animals can make a large number of tracks over a longer time period. It could simply be that the substrate was wet over a large area, for a long time. This is the geological perspective that pays greater attention to the dimension of time.

Figure 12.2B. *Part of the Salt Valley megatracksite, showing an area with intense dinoturbation.*

Several clues suggest that this latter perspective is the correct one; the track-bearing layer is associated with a major unconformity, a break in the regular sequence of the accumulation of strata. Sometimes unconformities result from the erosion of preexisting deposits, which removes part of a former accumulation, leaving a gap in the geological record; in other cases there is a gap because nothing was deposited during a particular period.

Imagine a lake basin steadily filling up with a rain of fine sediment that is settling out of suspension in the lake waters. This would result in a continuous sedimentary accumulation. If, on the other hand, the lake basin dried out, the settling of sediments would cease. It might be years, centuries, or millennia before the same process of accumulation would begin again. Although no sedimentary layers were deposited, there may be clues to the events that transpired during this gap in the history of sedimentary accumulation. The old lake bed may have changed color and chemical composition through oxidation, plant growth, or soil formation. All of these processes can alter the nature of the substrate without adding new layers, and, of course, tracks will leave an indelible mark.

The point is that if no new sedimentary layers accumulate, thousands of generations of tracks may be imprinted on this substrate. Even if the substrate (lake bed in this case) is dry for 360 days of the year, it only takes an occasional moistening to permit additional tracks to be made and added to the permanent geological record. To a biologist tracks may simply be a record of animal activity, but to a geologist the number of tracks or the density of dinoturbation is also a measure of the duration of an unconformity or a gap in the sedimentary record during phases of nonaccumulation. The principle is the same as that for using the thickness of sedimentary layers as a measure of time during phases of active accumulation. As stressed in Chapter 10, tracks accumulate when sediments are *not* being deposited, and vice versa.

It is interesting to note that all the tracks currently known from the top of the Entrada are three-toed varieties. Moreover, they all appear to be those of carnivorous theropods. This is consistent with the high proportion of carnivore tracks known from all the underlying desert deposits of Lower and Middle Jurassic age (Chapter 8). However, it is surprising, given the extremely large number of tracks, that no other types are represented. Sauropods, for example, are well represented in the overlying Morrison beds, but their tracks are still unknown at this level. Such differences appear to emphasize the strong environmental preferences of different trackmakers, for, as we have seen, sauropod tracks have yet to be found in desert sand deposits.

Moving forward in time to the 100-million-year-old beds of the Dakota Group, along Colorado's Front Range, we encounter an example of a Cretaceous megatracksite. Tracks have been known from these deposits for at least fifty years, but they were generally thought to occur at only a few isolated locations. In recent years, after a survey of the outcrops, a series of tracksites have been recognized from Boulder, Colorado, in the north to northeastern New Mexico in the south, a distance of several hundred kilometers. Where the track-bearing layers have been studied in detail, near Denver, Colorado, the zone of footprints appears to be almost continuous for at least 70 kilometers (Fig. 12.3).

Analysis of the deposits indicate that the tracks occur very near the top of the Dakota Group in a series of thinly bedded sandstones and shales usually interpreted as part of an extensive coastal plain environment. The track-bearing layers have been described variously as tidal flat, beach, levee, swamp, lake, and delta plain deposits. They are associated with layers rich in plant debris and valuable refractory clays[5] of the type associated with coal swamp deposits. The vegetation remains indicate a lush, relatively humid environment. The deposits show that sea level was relatively low and that a broad coastal plain existed.

As discussed in Chapter 8, the tracks represent a dinosaur community dominated by ornithopods with a variety of small to medium-size carnivores and a few birds. The most common ornithopod trackmaker is the type referred to as *Caririchnium*, with small hooflike front feet and broad three-toed hind feet (see Chapter 5). A few other tracks are similar to the classic *Iguanodon* tracks first described from England. The carnivorous dinosaurs are mainly represented by small to medium-size slender-toed trackmakers. These would have been gracile, ostrich-like dinosaurs of the type often referred to as coelurosaurs. There appears to be a distinct lack of large robust theropod trackmakers. Also conspicuous by their absence are sauropod tracks. As discussed above, brontosaur trackmakers are completely unrepresented in these well-vegetated, clastic coastal plain deposits throughout western North America and evidently worldwide. The purpose of restating this dinosaur footprint evidence is to stress that a specific dinosaur community was associated with this particular coastal plain. As we shall see, coastal plains elsewhere in the world at this time were home to very different dinosaur communities.

One of the better known and more accessible tracksites is situated west of Denver at a locality known as Alameda Parkway, or Dinosaur Ridge. Because *alameda* means "promenade," the track-bearing beds have been dubbed Dinosaur

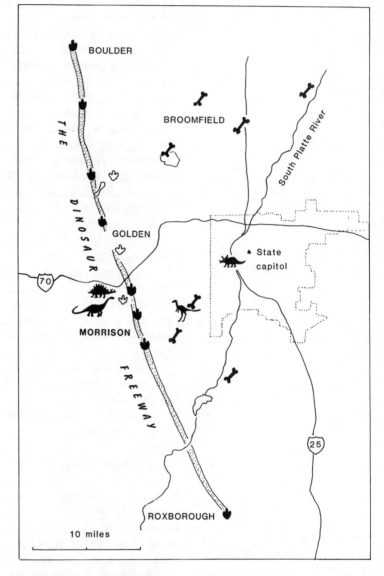

Figure 12.3. Distribution of track-sites in Dakota Group strata west of Denver shows that tracks can be found in this deposit over a distance of at least 65 kilometers. This has given rise to the name Dinosaur Freeway. Other track and bone sites of different ages are also shown with appropriate symbols.

Promenade. Following the construction of a new freeway parallel to the track-bearing outcrops of the Front Range, the Dinosaur Promenade also became known as the Dinosaur Freeway (Fig. 12.3).[6] The number of the modern thoroughfare is C-470 (Colorado 470); the ancient route is Cretaceous-470.

The designation of a freeway or promenade is appropriate. We know that these coastal plain deposits formed part of an extensive north–south trending shoreline along the western shore of an inland seaway that extended northward from the region of the present-day Gulf of Mexico. Such low-lying,

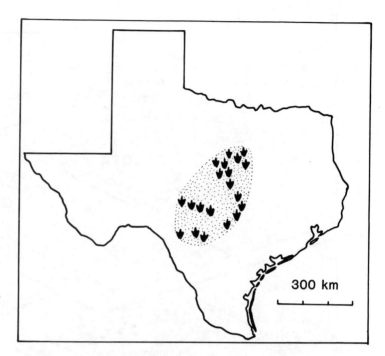

Figure 12.4. *The known extent of the Upper Glen Rose megatracksite complex of Texas. The area is on the order of 100,000 square kilometers.*

300 km

vegetation-rich environments were probably prime habitats for herbivorous ornithopods.[7] When they were traveling or migrating in search of food, it is reasonable to assume, they followed the general shoreline trend, at least on occasions. Some have expressed the possibility of shore-parallel migration routes. Although there is no absolute proof that Colorado's Dinosaur Freeway was a migration route, the similarity of track types over a wide area from northern Colorado to northern New Mexico increases the likelihood.

This similarity, however, is only evident where the deposits remain similar, that is, in the sandstones and shales of the Dakota Group. Further to the south, in limestone-rich deposits of about the same age, the footprint assemblages are entirely different. These southern deposits are the famous track-bearing layers of Texas. In a recent study, Jeffrey Pittman noted that the majority of the more than forty known Texas tracksites occur in two distinct zones. The lower of these zones, at the top of the Lower Glen Rose formation, includes the Paluxy River sites in the Dinosaur State Park area and the Mayan Ranch site, and the upper zone, at the top of the Upper Glen Rose, includes the Davenport Ranch site.[8] Both zones extend for several hundred kilometers laterally and over areas on the order of several thousand square kilometers (Fig. 12.4). As far as is currently known the upper zone exhibits a greater number of exposed tracksites (Fig 12.5), allowing for a more con-

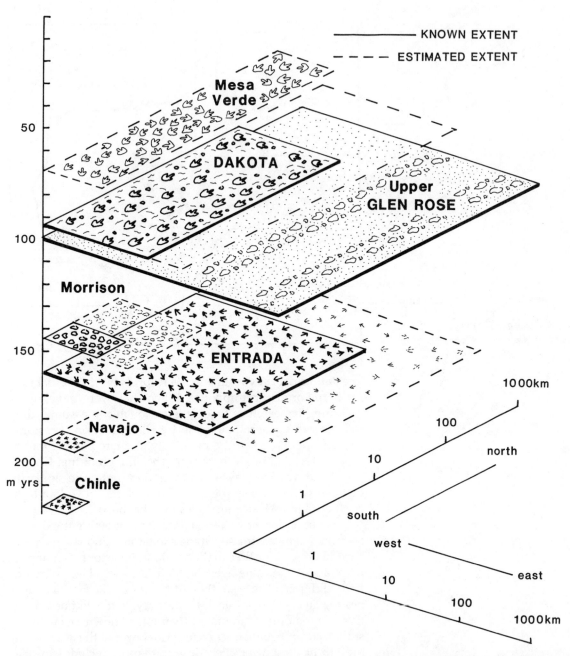

KNOWN EXTENT

ESTIMATED EXTENT

Mesa Verde

DAKOTA

Upper GLEN ROSE

Morrison

ENTRADA

Navajo

Chinle

50

100

150

200

m yrs

1000km

100

10

1

north

south

west

east

1

10

100

1000km

Figure 12.5. *The known and esti-mated lateral extent of the Entrada, Glen Rose, and Dakota megatracksites. The estimated extent of the latest Cretaceous Mesa Verde track beds are also indicated, as are some smaller track-sites.*

tinuous representation than can be demonstrated in the lower zone. Pittman has found many new sites just by following or tracing the tracksite layers of strata to new locations. This means that dinosaur trackers are using predictive techniques, one of the hallmarks of the scientific approach.

Pittman's studies have revealed that, like the other mega-tracksite examples, the Texas track zones are associated with

low-lying, coastal plain. The Texas coastal plain was the ancient Cretaceous Gulf of Mexico, characterized by limestones
of the type referred to as carbonate platform deposits. Pittman also notes that the track zone probably represents an
expanding coastal plain associated with a period of relatively
low sea level. In this paleoenvironmental setting the predominant trackmakers appear to have been carnivores and brontosaurs, with very few large ornithopods. As discussed in
Chapter 8, this is a good example of a sauropod–theropod
community. Again, the evidence demonstrates how this particular coastal plain has its own distinct dinosaur community:
three different megatracksites, all associated with low-gradient
coastal plain settings, but all frequented by quite distinct dinosaur communities.

The only other known coastal plain carbonate platform environment with similar track assemblages that could possibly
be considered a megatracksite complex is the late Middle Jurassic sequence of deposits from North Africa. These are not
only rich in sauropod and theropod tracks, but they occur in
deposits that are about the same age as the Entrada beds in
Utah. This suggests that a late Middle Jurassic megatracksite
in sandy coastal plain deposits in western North America can
be matched with a limy coastal plain (carbonate platform)
megatracksite of about the same age in North Africa. Similarly, the sandy coastal plain megatracksite complex of Colorado and New Mexico almost matches with the megazone(s)
in the carbonate platform deposits of Texas.[9]

These preliminary conclusions need to be tested. Nevertheless they indicate that entirely different footprint assemblages
from different paleoenvironments have geological connec-
tions related to worldwide fluctuations in sea level.

It makes sedimentological sense that tracks will form and
be preserved after sea level drops and coastal plains expand.
At this time rivers cut new valleys to reach the sea at its new,
low level. As sea level ceases to drop, however, the valleys
start to fill with layers of sediment. Tracks can be preserved
in this situation, as in the Dakota example. Sometimes, however, rising sea levels cause erosion and destroy sediments
and the tracks they contain.

A worldwide drop in sea level is like a large-scale or prolonged ebb of the tide or shrinking of a lake. Flora and fauna
can encroach into newly exposed areas. In effect, the tracks
are evidence of the dinosaurs' response to the development
of new coastal plain habitats. Given the many thousands to
millions of years involved in the broadening, or expansion,
and subsequent shrinking of these coastal plains, it would be
unthinkable to suppose that vegetation and dinosaurs did not
spread outward to take advantage of these extensive new

habitats. The footprint evidence proves that they did frequent these areas.

It has been suggested by Jack Horner and others that the expanding and contracting of coastal plain habitats significantly affected dinosaurian evolution. Horner cites Robert Bakker as suggesting that rises in sea level and contraction of the coastal plain "caused numerous extinctions and rapid speciation." Put another way, "the growing sea wiped out habitat and squeezed everybody into the mountains."[10] As pointed out above, there is not always a good record of the rising sea level phase (transgression). However, when sea level dropped (regression) and habitats expanded, "vast new territories, new ecological niches, were opened up to be colonized by opportunistic species." These appear to be the times when some megatracksites show up in the geological record. Even if the tracks were made and preserved during the rising sea level phase (transgression), as evidence suggests for some sites, this circumstance requires an earlier sea level drop (regression) to establish the broad coastal plain in the first place.

If Horner is right and there were rapid radiations of new dinosaur species into these new territories, we should be aware that the tracks represent dinosaurs in these particular expansive evolutionary circumstances. The abundant carnivore tracks in the Entrada deposits of Utah may represent an evolutionary radiation among theropods, and the proliferation of ornithopod tracks in the Dakota may indicate comparable expansions for these trackmakers. Similarly, we may discover clues to the evolution of sauropods in both the Jurassic and the Cretaceous by understanding the track records in North Africa and Texas.

The discovery of megatracksites has heralded a new era in dinosaur tracking. It is an unprecedented step for dinosaur trackers to consider applying their evidence to global questions like worldwide sea level fluctuations. On a less sophisticated geological level, the sheer abundance of new evidence is staggering, several orders of magnitude greater than we had ever suspected. Literally millions of footprints exist in a single square kilometer within these megatracksites, and the sites extend over hundreds to thousands of square kilometers. This means that motorists in cities like Denver, Colorado, and Dallas and Austin, Texas, drive over millions of tracks every day as they commute to and from work above the subsurface tracksite layers. For the scientist these megatracksites provide data for decades of future research.

The megatracksite discoveries raise as many questions as they answer, but this is the nature of science. Unless we wish to venture into the realm of wild speculation, this is as far as we dare take our analysis, with our present knowledge. Many ideas and hypotheses will need to be tested. Megatracksites require further study, but now at least we know where to look for them and can explore their extent using scientific prediction methods rather than guesswork.

This is an appropriate place to end the scientific part of the book. We can step back and take a reflective look at the science of dinosaur tracking and what has been learned over the decades and particularly in recent years. Our next chapter will deal with what can happen if we let unbridled imagination, wild speculation, or sheer laziness replace a commonsense, objective, and scientific approach. As we shall see, there is a need to discriminate between science fact and science fiction. Dinosaur tracking has come too far in recent years to be reduced to myth and fantasy.

13

Myths and misconceptions

Reconstruction of a carnivore attacking a sauropod, based on controversial evidence from Texas. Artwork by Nola Montgomery, courtesy of Texas Parks and Wildlife Department.

We are not authorized in the infancy of our science, to recur to extraordinary agents.

– Charles Lyell

In this chapter we enter the arena where anecdote comes face to face with skeptical science. The results can often be amusing, but we seek more than the fun of exposing the ridiculous myths that have crept into science. The main objective of this chapter is a plea for caution. Rather than speculating wildly, we must patiently accumulate good evidence and critically evaluate our hypotheses regarding tracks. We must learn to walk before we try to run.

Although the 1980s was a revolutionary decade for dinosaur tracking, many misconceptions have arisen from a poor understanding of the dinosaur footprint evidence. This is not too surprising because dinosaurs are like mythological creatures to the popular mind. If it were not for the tangible evidence of their fossilized skeletons, they would be as incredible as dragons.

Since tracks furnish some evidence of dinosaur behavior, some authors have let their imaginations run amuck. They have inferred all manner of behavior, from unusual waddling and hopping to swimming, predator–prey skirmishes, stampedes, and social structuring in herds. Unfortunately, however, the majority of evidence indicates generally pedestrian activity, rather than acrobatic feats. Ninety-nine percent or more of all trackways reveal walking. Reported hopping and other unusual forms of locomotion often turn out to have much more mundane explanations.

Dinosaurs that dither, dawdle, and deviate

There are a number of instances of dinosaur trackways intersecting in seemingly unusual configurations. There is one slab from Dorset, England, for example, on which a dinosaur trackway heads off the edge in one direction at the same point that another one of about the same size crosses at 90 degrees. The configuration looks like a dinosaur turning abruptly to the right, but it is only *a part of the picture*. There are countless other sites where trackways intersect at different angles. At first sight such crossroads appear to reveal evidence of dinosaurs doubling back, deviating, dithering, or dawdling. I've often been at sites where visitors interpret complex track configurations as evidence of dinosaurs slowing down, speeding up, or making abrupt turns. Usually these complex interpretations result from a misunderstanding of which tracks belong to which trackways. Confusing configurations can sometimes arise when dinosaurs traveled from soft to firm ground; their trackways appear and disappear abruptly, sometimes resembling the pattern of a sharp turn, a dither, or a dawdle.

There are a couple of valid examples of abrupt changes in direction, discussed in Chapter 6. We presented no interpretation of the activity there because, unfortunately, no motive can be discerned from the evidence. Why does the unusual carnivore trackway reported by Phil Currie from the Minnes Formation of Canada record an abrupt 90-degree turn to the right? Why does a brontosaur trackway from the Morrison Formation of Utah show a right turn in the direction of what is now Colorado?[1] Why did these animals not turn to the left? Perhaps they were influenced by the clockwise Coriollis force, which makes moving objects veer to the right in the Northern Hemisphere[2] and possibly affects the migration patterns of wild animals. This intriguing explanation is neither plausible nor convincing. Unfortunately, it is the sort of unfounded speculation that sometimes get confused with scientific hypothesis, and once it is put in writing, someone somewhere will believe it. If a fallen tree or some other recognizable obstacle were preserved in the path of these animals, we would have grounds to speculate on the reason for the change in direction. But lacking these, we cannot use footprints to second-guess motives.

The megalosaur that didn't waddle

It is a common misconception that all dinosaurs were cumbersome, that, like cold-blooded crocodiles and alligators, they had a sprawling gait. Trackway evidence is proving that they did, in fact, walk erect most of the time, but it can sometimes be misleading.

In 1962 some new *Iguanodon* footprints were reported from the Purbeck limestone of Dorset. At first sight they appeared to show a wide trackway of an *Iguanodon*. The discovery was cited as evidence of an animal walking with its left and right hind feet widely splayed; it was reported in *New Scientist* magazine as "the slow march of a Purbeck *Iguanodon*" an "animal that appears to have moved very ponderously indeed."[3] Some researchers suggested that it had been moving uphill. However, further work revealed that the supposed single trackway was in fact two separate trackways attributable to two animals progressing in the same direction. The proof came from a third trackway with the same orientation but a very narrow configuration. During the reevaluation it also became clear that the tracks were probably attributable to a *Megalosaurus*-like dinosaur, *not* to an *Iguanodon*. Thus the British track-

Figure 13.1. *A trackway interpreted as that of a hopping dinosaur was later reevaluated as that of a swimming turtle. Artwork by Edward Von Mueller.*

ing sleuths solved the case of the megalosaur that did not waddle.

The myth of the hopping dinosaur; or, did dinosaurs play hopscotch?

In 1984 a group of French workers reported on the tracks of a Jurassic animal that appeared to have hopped across a limy seafloor substrate. They attributed the footprints to a theropod and, to explain its tracks in marine sediments, concluded that sea level dropped periodically to expose the limy seabed. They named the trackway *Saltosauripus*, meaning "hopping reptile tracks."[4] Not only did the tracks imply an unusual gait for a dinosaur, but they also suggested land where geologists had always inferred a marine environment.

Recently Tony Thulborn reevaluated the evidence by comparing the trackway with those of known hoppers like the Australian kangaroo, its relatives, and certain birds. He found that the French tracks, set widely apart, are quite unlike those of modern hoppers. The Jurassic trackmaker would have had to have been progressing in a permanent wide straddle, like a fully extended hopscotch player jumping on consecutive double rows. It just did not make much sense to Thulborn, and rightly so. Instead, he proposed that the trackmaker was probably a wide-bodied animal, like a turtle, swimming in shallow water and touching the bottom with synchronous strokes of its paddle-like feet (Fig. 13.1).[5]

Thulborn is surely essentially correct. Turtle tracks are reported from these sediments, whereas other dinosaur tracks are not. The turtle may have been swimming in shallow water or even underwater. Either way, the turtle interpretation requires less stretching of the imagination. Common sense and the principle of Occam's Razor requires that the "simplest of competing theories be preferred to the more complex."[6] It is more probable that turtle left tracks in their known habitat than that the sea level suddenly dropped allowing dinosaurs to play hopscotch on the exposed continental shelf.

Trackers and attackers

Do the tracks of a theropod dinosaur from Texas indicate that it attacked a brontosaur? Many people say the answer is yes. The trackway evidence shows a brontosaur trackway that runs parallel to the supposed attacker before both trackways converge. The American Museum of Natural History used segments of these two trackways to mount an exhibit showing an allosaur stalking a brontosaur. Some people believe that the brontosaur trackway ends at the point where it converges with the theropod trackway, implying that the luckless sauropod was finished off by its attacker. Others believe that evidence for the coup de grace lies a few paces further on, where the trackways disappear under the river bank.

As discussed elsewhere, this discovery of converging trackways, made by Roland T. Bird, has led to one of the most controversial debates generated by fossil footprint evidence. The evidence, which originates from Dinosaur State Park on the Paluxy River in Texas, consists of twelve brontosaur trackways all heading in the same direction, "followed" by at least three carnosaur trackways (see Chapter 7). We know the carnosaurs passed later because their tracks overlap those of the brontosaurs. In one place where a carnosaur trackway crosses or overlaps a brontosaur trackway, one of the theropod tracks appears to be missing. Bird inferred that the theropod must have attacked the sauropod's left rear flank, biting or grabbing on so that the bold attacker was bodily carried by its victim for a short distance before putting its feet down again to resume normal progression under its own steam (Fig. 13.2).

As outlined earlier, this scenario is fanciful at best and probably wrong. Nevertheless, the Dinosaur Valley State Park brochure reports a "remarkable set of double tracks left by a giant sauropod dinosaur followed by a large carnivorous dinosaur . . . [an] impressive record of an ancient hunt." Such is the power of the graphic reconstructions of this scene that appeared in Bird's book and as cover illustrations on several recent field guides.[7] However, both the sauropods and the theropod trackways show little or no deviation or obvious

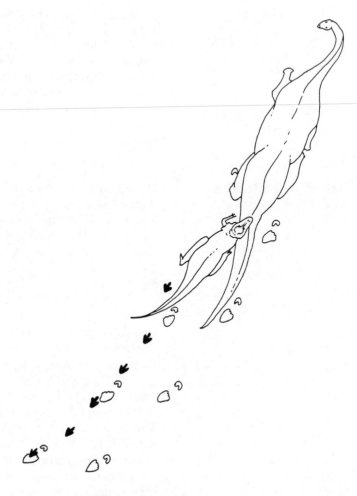

Figure 13.2. Roland T. Bird's scenario of a carnosaur attacking a sauropod. Based on footprint evidence from Dinosaur State Park, Texas.

change of pace. The theropod also crosses the sauropod's trackway from left to right within a few steps of the inferred contact point. This would be almost impossible in a rear flank attack, without the sauropod's long tail getting in the way or requiring the theropod's detaching itself and waiting while the sauropod walked calmly away. As a more probable scenario, the theropods followed the sauropods at some distance, and when one of them trod on the recently compacted sediment in a brontosaur track, it left no impression. This means, of course, that the famous American Museum exhibit is misleading. A small pack of theropods may have been stalking a herd of large brontosaurs, but there is no evidence for a one-on-one attack.

A widely used textbook has adopted a dinosaur footprint interpretation exercise based on this supposed skirmish.[8] The exercise purports to allow students to work with "sketches of a set of fossil footprints made by animals more than 100 mil-

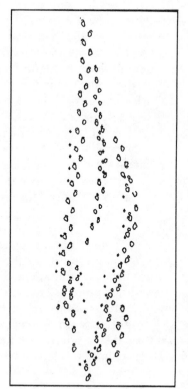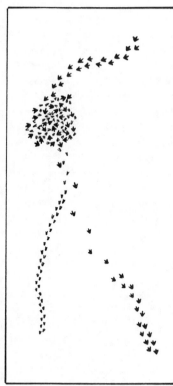

Figure 13.3. *A comparison of Bird's map of the Dinosaur State Park site (left) and the site map concocted by textbook authors for a high school general science exercise (right).*

lion years ago." As we can see from Figure 13.3, the exercise is fun and educational to a degree, but it is pure fantasy. The made-up tracks show two animals converging. As they draw near, the predator speeds up and attacks the other animal, its poor victim. They run around in circles, and the victim succumbs, perhaps to dizziness. Then the predator walks off, apparently with the victim in its jaws or its stomach. Comparing Bird's map with the map concocted for the exercise, however, we see that there is virtually no similarity. The bright student may question the intelligence of a prey species that does not speed up and try to run away until the last minute, but instead runs in ever-decreasing circles! The exercise has its merits; it stimulates imagination and may help students formulate multiple working hypotheses. But nowhere does its author admit that the scenario is purely imaginary. No such real fossil footprint evidence exists.

There is a further strange irony in the attack scenario: The American Museum dinosaurs mounted above these Cretaceous tracks are from the Jurassic period. When Bird visited the Purgatoire site in Colorado on his way to Texas, he saw genuine brontosaur tracks in the Jurassic Morrison Formation, but because they were not as crisp as the Cretaceous tracks he was soon to uncover in Texas, they were all but forgotten. However, when it came to finding a dinosaur to mount above the footprints, the museum had little choice but using Morrison dinosaurs from the Rocky Mountain Region. The museum could find no suitable Cretaceous skeletons available. At the time, and on many subsequent occasions, the incongruity of this display has been commented upon. Few if any scientists would endorse an exhibit with two or more dinosaurs of different ages side by side in the same scene, but the spectacle of 150-million-year-old dinosaurs making 100-million-year-old tracks has survived for decades.[9] (Other museum displays also mix bones and tracks of different ages. At the Natural History Museum in Albuquerque, New Mexico, another Morrison sauropod is reconstructed above Cretaceous tracks. These giant steps forward for Morrison dinosaurs are not monumental strides for science.)

We should not leave the reader with the impression that dinosaurs never attacked each other or that the track record might not potentially reveal evidence of an attack. During a recent study of modern tracks, we found dramatic evidence of a fight between two hippos (see Fig. 13.4). We had staked out an area of mudflat on the north shore of Lake Manyara in Tanzania in order to survey it every day for new tracks.[10] On the night of June 3–4, 1989, two battling hippos ran into the area from the west. They were obviously fighting, because at intervals there were long slide and scuffle marks, accompanied by pools of blood, and changes in the directions of their trackways. After two bouts of running and scuffling (Fig. 13.4), the rivals separated and faced each other. At this point we could tell that the smaller hippo was the one that was bleeding. Not surprisingly, the larger hippo had been the one to establish the upper hand and drive its smaller rival away from the lake. As the vanquished underdog retreated, the triumphant victor strategically splattered a monstrous dung deposit on a prominent tuft of grass, a unmistakable territorial marker and monument to victory.

Hippos are notoriously pugnacious, and territorial battles are commonplace. There could be little doubt as to the exact significance of the several hundred yards of variable trackways we studied. We also found the trackways very educational from a paleontological perspective. They showed us that an attack reveals evidence of running, slipping and sliding, and sudden changes in direction. We would not expect to find blood and dung preserved in the fossil record, but the trackways would presumably be irregular, unusual, and significantly different from those exposed at Dinosaur State Park in Texas.

A REAL ATTACK

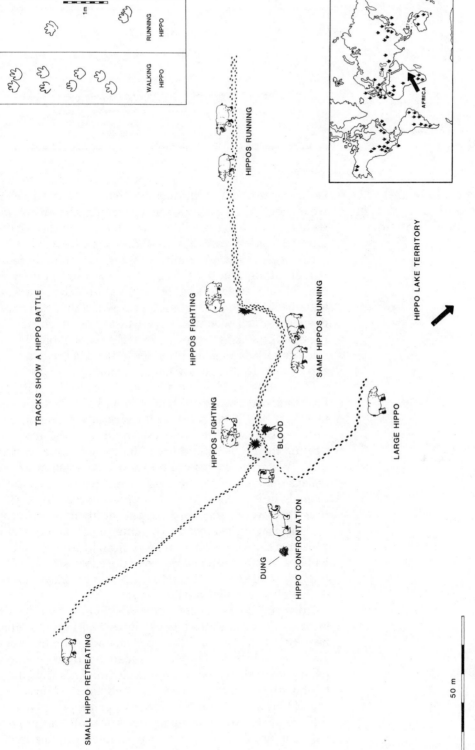

TRACKS SHOW A HIPPO BATTLE

WALKING HIPPO

RUNNING HIPPO

1 m

AFRICA

HIPPOS RUNNING

HIPPOS FIGHTING

HIPPOS FIGHTING

SAME HIPPOS RUNNING

BLOOD

LARGE HIPPO

HIPPO LAKE TERRITORY

DUNG

HIPPO CONFRONTATION

SMALL HIPPO RETREATING

50 m

Figure 13.4. Schematic map of track-ways showing a hippo battle at Lake Manyara in Tanzania (1989). Note running, scuffling, confrontation, and separation phases.

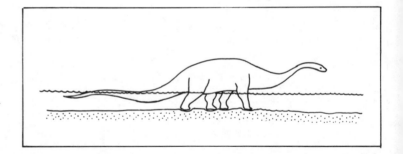

Figure 13.5. Roland T. Bird's floating tail scenario to explain why tail drag traces are not found at the Davenport Ranch site.

Leave them alone and they'll come home, *floating* their tails behind them

Did dinosaurs drag their tails? As discussed in Chapter 6, if we rely on tail drag marks, the answer in 99.9 percent of cases would be no. Tail drag traces are very rare, and most of their reported cases are ambiguous or dubious at best.[11] This makes sense because, energetically speaking, it is difficult to drag a tail, and tough on the tail as well. Few animals do it. However, in some instances lightly impressed tail drag marks might have been made on surface layers while the tracks sank in deeper; the tail would not exert nearly as much downward pressure, or sink as deeply, as a weight-bearing foot. The track–undertrack relationship provides a plausible, but unproved, explanation for the lack of tail drag marks.

Other, less credible explanations have also been proposed. Following his discovery of sauropod tracks at Davenport Ranch in Texas, Bird proposed that the lack of tail drag marks implied that the animals were wading through shallow water, floating their tails behind them. Bird's reasoning was convoluted: Water shallow enough to allow the small sauropods to make a normal walking trackway would have offered the larger animals little or no buoyant support, except for their tails; thus they were capable of supporting their weight on land without leaving tail traces. This interpretation led Bird to reconstruct the sauropod in the American Museum exhibit with its tail about 15 centimeters above the substrate, as if floating (Fig. 13.5). He had planned to put a sheet of glass at the level of the tail to simulate shallow water.

However, Bird was almost certainly wrong about the floating tails. The same trackways show "mud lumps" that fell from the animals' feet; these are characteristic of trackways made by animals walking over substrates that are emergent, not submerged. Nevertheless, despite the improbability of the tail-floating scenario, at least one study of the Moroccan sauropod tracks adopted Bird's idea as a possible explanation for the lack of tail traces.[12] Again, we encounter an example of failure to use Occam's Razor: The simplest explanation – that they did not drag their tails – was never considered.

The structured herd hypothesis

Figure 13.6. Herd structure is a complex phenomenon whose interpretation is based on the overlapping relationships of tracks and the sequence in which they were made. Each circle represents one of the twenty-three Davenport Ranch sauropod trackmakers. Circle size is proportional to track size, and arrows point toward tracks they overlap. Circles to the left represent animals that passed by earliest, and circles to the right represent those that crossed the area last.

We have noted that the Davenport Ranch tracksite provided footprint evidence of gregarious behavior, or herding, among dinosaurs. According to Robert Bakker the distribution of large and small trackways indicates a "structured" herd, with the "very largest footprints made only at the periphery of the herd; the very smallest . . . only in the center."[13] John Ostrom questioned this interpretation, suggesting that the track evidence could not support such inferences,[14] but no one analyzed the footprint data in detail.

When carefully studied, the tracks reveal something of the sequence of events that took place.[15] From the overlapping relationships of the trackways, we can work out which animals led the way and which followed (Fig. 13.6). The evidence indicates that it was the larger animals that tended to lead the way, with the smaller ones behind.[16] The herd was moving at about 1 to 2 meters a second, a modest walk. It

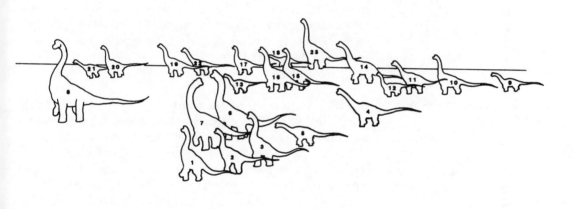

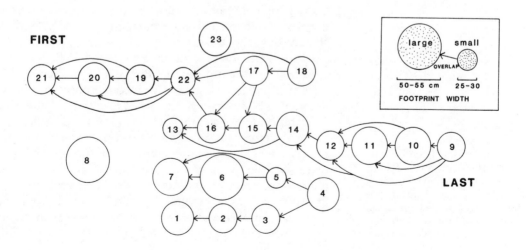

took a minimum of 30 to 60 seconds to cross the area where the tracks are preserved. We can also determine that the herd was generally veering from right to left. Moreover, it was strung out in a line. John Ostrom is right in concluding that there is no evidence for the type of herd structure proposed by Bakker. However, this does not mean that brontosaurs never encircled their young to protect them. Jurassic trackway evidence from the Purgatoire site indicates a group of small brontosaurs flanked by a group of larger individuals.[17] However, as discussed in Chapter 7, all manner of herd configurations are possible, and the protective or structured herd formation is not proved.

The myth of the structured herd persists.* Recently *Natural History* magazine published a statement about duckbill dinosaurs protecting their young at the center of a herd (see Chapter 7), and *National Geographic* magazine published a superb poster of a herd of giant *Ultrasaurus* protecting their young within the center of the herd as they are being attacked by marauding carnivores.[18] This latter is a combination of the attack scenario and the structured herd hypothesis, with protective *Ultrasaurus* adults thrown in for good measure. Three myths for the price of one.

Did brontosaurs swim out to sea?

Based on evidence from the third of his famous Texas tracksites, the Mayan Ranch locality, Bird proposed that brontosaurs were aquatic.[19] He analyzed a single trackway that consisted largely of shallow front-foot impressions, attributing the lack of hind footprints to the brontosaur's being buoyed up in about 3 meters of water, only touching bottom with its front feet. Bird's interpretation implies that brontosaurs behaved somewhat like large seals, which venture out from the shoreline into the shallow marine environment. Although many people do not unreservedly accept this interpretation, they have been led to believe that footprint evidence supports at least some aquatic activity by brontosaurs. But do the tracks support the notion of swimming sauropods?

At first sight it appears that Bird was right. Recent discoveries by Shinobou Ishigaki of Moroccan brontosaur trackways dominated by front footprints seem to suggest that such configurations are not so rare.[20] However, if we consider preservational factors and distinguish true tracks from undertracks, we get a whole different perspective on the swimming sauropod scenario. Most of the well-known sauropod tracks are quite deep, up to 25 centimeters at the Paluxy River site and in excess of 5 to 10 centimeters in a large number of documented examples; often the tracks are deeper than the thickness of the beds they penetrate, leading inevitably to under-

*As part of its "Land Before Time" promotion, Pizza Hut developed an educational package for students in grades 1 through 6. The package showed a fictitious map of adult brontosaur trackways flanking those of juveniles.

Figure 13.7. *A comparison of Bird's swimming sauropod interpretation (above) and the undertrack interpretation (below). Note the hind footprint (dotted line) missed by Bird. Artwork by Edward Von Mueller.*

tracks on underlying beds. The Mayan Ranch and Moroccan "swimming" sauropod trackways are remarkably shallow and lack well-defined outlines, both typical characteristics of underprints. Moreover, Bird missed one of the hind footprints (Fig. 13.7), and the spacing of the front footprints is typical of gaits seen in normal progression on land, so we see that the swimming hypothesis may be flawed.[21]

Essentially what has been overlooked in the previous studies is the fact that a thin layer of sediment, say 2 to 5 centimeters, can buoy up or support an animal just as effectively as several meters of water. The underprints made on buried layers below this thin layer may be shallow and incomplete.

Toe or claw traces may be seen where the foot's protruding extremities penetrated most deeply. Because the rest of the track is not visible, such partial footprint traces misleadingly give the impression of an animal touching the substrate only with the tips of its toes.

Including the possibility of undertracks, we find swimming dinosaur scenarios improbable, at least in a majority of cases, and I suspect that in most cases they are wrong. Incomplete footprints have been cited as evidence of swimming dinosaurs in several other studies, but an undertrack interpretation was not even considered in most of them. Among the groups supposedly displaying aquatic behavior the literature cites Lower Jurassic theropods and protoiguanodontids, Middle Jurassic and Early Cretaceous brontosaurs, and Early Cretaceous ornithopods and ornithischians.[22] What is at issue here is the reputation of ichnology. There is little doubt that dinosaurs, like most animals, were capable of swimming, but suggesting in the face of the strong evidence for terrestrial adaptations by brontosaurs and other dinosaurs that tracks argue for aquatic behavior is detrimental to the science. Trackers must learn to distinguish between tracks and undertracks. With a few exceptions, the studies that cite tracks as evidence of swimming dinosaurs are probably seriously flawed.

The pterosaurs that never landed

Although pterosaurs are not dinosaurs, they are close relatives (class Archosauria), and their remains occur in many of the same formations that yield dinosaur bones. In 1957 William Stokes reported supposed pterosaur tracks in the famous dinosaur-rich Morrison Formation.[23] Later he and Jim Madsen reported similar tracks, also attributed to pterosaurs, from the Navajo Formation.[24] Because these tracks were named *Pteraichnus* (= *Pteradactyl* traces), the notion that they were pterosaurian in origin became firmly entrenched, and several other similar tracks were attributed to these airborne reptiles. In the 1980s Kevin Padian, a pterosaur expert, began to doubt that these tracks could be attributed to pterosaurs walking on land. He teamed up with Paul Olsen, another track expert, to examine the tracks in detail and compare them with those one would expect pterosaurs and other archosaurs to make. Their results, reported in 1984, were that the tracks had probably been made, not by pterosaurs, but by other archosaurs, namely crocodilians.[25] They backed up this conclusion by showing the similarity between the fossil tracks and those made by small modern crocodiles. Further studies by the author suggest that none of the tracks known from the Navajo Formation were made by pterosaurs.[26]

These conclusions mean that no proven pterosaurian tracks

are presently known. This does not suggest that pterosaurs never landed on wet or soft ground where they might have left tracks, but that they did so relatively rarely or in areas that we have yet to discover. Obviously airborne creatures do not leave tracks while flying; neither do they leave tracks on firm substrates like branches, rocks, or high ground, where they may choose to alight or perch. Despite this cautionary tale, other recently discovered crocodile-like tracks have been interpreted as pterosaurian.[27] Interesting ideas will not go away, even if they are improbable or largely disproved.

Dinosaurs and dynamite

Most layers of strata are planar, flat and boardlike, with a limited number of indentations. Features like ripple marks or mud cracks are fairly obvious, and in most case so are footprints. However, sometimes regular or irregular holes are hard to interpret. In my experience a minority, perhaps 1 to 5 percent, of the tracks reported, turn out to be holes or indentations that originated as a result of erosion, water action, artificial excavation, or blasting. In most cases such unusual or artificial features are not confused with dinosaur tracks or documented as such in the scientific literature. However, in some localities genuine fossil tracks do occur in close proximity to such unusual features, or pseudotracks, leading to confusion.

Along Alameda Parkway, west of Denver, at the Dinosaur Ridge locality dubbed the Dinosaur Freeway, there are a number of blast holes found in the same general area as layers with dinosaur tracks. Blast holes are not uncommon along highways where rock has been removed to accommodate road building. Moreover, because construction crews often place their charges at regular intervals along a rock outcrop, the blast holes are often regularly spaced, rather like a trackway. Geologists can usually recognize blast holes because of a distinctive pattern of radial fractures not normally seen in tracks or any other sedimentary rock features. However, it is not always easy to distinguish such features. Old blast holes may be weathered and look different from fresh ones. Moreover, some genuine tracks are associated with radial cracks[28] or occur in strata that are thoroughly cracked or fractured as a result of natural geological deformation. It is always worth looking into reports of unusual indentations to determine if the cause was dinosaurs or dynamite.

The crocodile crusher

In the 1920s and 1930s the discovery of giant hadrosaur tracks in the Cretaceous coal mines of Colorado and Utah generated much excitement. Barnum Brown of the American Museum

Figure 13.8. Workers excavate a large track-bearing slab from a coal mine in western Colorado. (Detail of trackway below.) Artwork by Edward Von Mueller.

played down the obvious duck-bill affinity of the tracks[29] and stressed their gigantic size, suggesting that no known trackmaker was large enough to make such huge footprints. He dubbed the trackmaker the Mystery Dinosaur and claimed that it was capable of covering 15 feet (almost 5 meters) in a single step (Fig. 13.8). To support this claim he mounted a major excavation from a coal mine in Cedaredge, Colorado, to remove two large tracks that he claimed showed this monstrous 15-foot step. Although it was later demonstrated that this "step" was in fact a stride of two consecutive steps of $7\frac{1}{2}$ feet, speculation about the giant mystery trackmaker remained rife.

Brown's emphasis on giant tracks and giant steps fostered a *Guinness Book of Records* attitude toward footprints. People were eager to claim record-breaking statistics, and soon a step of 16 feet 3 inches was reported.[30] If this was not phenomenal enough, the same author had earlier reported a site where a dinosaur had stepped on an animal resembling a crocodile. Unfortunately no other information was given.[31] Although it is theoretically possible that a dinosaur could have stepped on a live crocodile, or one that was already dead, the account lacks scientific substance and must reside in the files under the heading Crushed Crocodile Myth. For those interested in unusual paleontological phenomena, there is an example of a fossil flamingo that was stepped on by a camel during the Tertiary period.[32] Also, of course, as shown in Chapter 11, dinosaurs did trample clams.

The sprinting duckbill

Although scientists never took the crushed crocodile story seriously, considerable attention was devoted to the giant steps of the Mystery Dinosaur. In the late 1970s and early 1980s there was much debate over the speeds reached by dinosaurs. Brown never made such claims, but other researchers, relying only on his published account, interpreted the long steps as evidence for running.[33] The giant slab excavated by Brown, actually shows a partial track midway between the two purported to be a single 15-foot step. This piece of evidence, pointed out by Thulborn,[34] obviously changes the picture considerably. The animal was not sprinting along at high speed. Its steps and hence its rate of progression were only half as dramatic as Brown and the high-speed advocates would like to believe. Again, not such an exciting scenario, but realistic and compatible with the actual footprint evidence.

Mesozoic human tracks

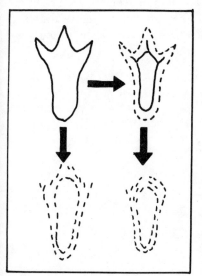

Readers may be aware of claims that human footprints have been discovered along side those of dinosaurs. As mentioned in Chapter 2, these tracks are elongate, with heel-like impressions that look superficially human. The problem, of course, is that they occur in rocks about 100 million years old, primarily in the Cretaceous deposits of Texas. Moreover, most of these so-called man tracks are much too big to be made by normal humans; at 50 centimeters in length they definitely fall in the Big Foot category.

Many of the man tracks claims have been made by the so-called creationists in an attempt to cast doubt on evolutionary theory and the well-established geological time scale. Although these claims are not taken seriously by seasoned geologists and paleontologists, they do cause consternation and confusion among laypersons without an earth science background. And as in an earlier example, they can result in erroneous or misleading contents in basic textbooks. Consequently it is encouraging that the confusion about these so called man tracks has recently been resolved by a careful study. In an attempt to understand their origin and significance, Glen Kuban undertook a thorough study of these elongate indentations and concluded that they were made by dinosaurs.[35]

Figure 13.9. *So-called man tracks are actually attributable to dinosaurs (after the work of Glen Kuban, 1989).*

He even showed that some tracks had three quite distinct toes (Fig. 13.9). They were made by carnivorous dinosaurs that were leaving metatarsal impressions. As we have seen, plantigrade progression, although uncommon among dinosaurs, was not unknown. As a result of Kuban's work, the Creationists withdrew many of their antievolution claims, particularly those based on footprint evidence. Clearly a victory for science over myth.

Few footprints: bone bonanzas

A final myth involves the perception that footprints are rare and rarely associated with bones. By now the reader should be convinced that dinosaur tracks are common, even abundant, in Mesozoic deposits worldwide. While it is true that footprints and bones sometimes occur separately, they are often found together. As discussed in Chapter 8, tracks are quite common in several deposits that have yielded bone bonanzas. The Chinle, Morrison, and Wealden deposits provide well-known Triassic, Jurassic, and Cretaceous examples. The generalization that footprint sites are few and far between is no longer acceptable.

Conclusions

Many myths have arisen from ignorance leading to incorrect or fanciful interpretation of dinosaur tracks. Although it is easy to discredit these myths, once an intriguing idea is introduced into the literature, it tends to stick, even after it has been decisively debunked. The lesson seems to be, keep an open mind and entertain all interpretations before drawing conclusions. Use Occam's Razor, multiple working hypotheses, and, above all, common sense.

14

The dinosaur trackers

Small Grallator *tracks and large* Eu-brontes *footprints were among the first fossil tracks ever named and studied. They are typical of Early Jurassic dinosaur communities and have been found in many regions of the world. Artwork by Gregory Paul,* Predatory Dinosaurs of the World, *N.Y. Simon & Schuster, 1988.*

The feet of animals furnish, probably, the best means . . . of judging the class, family, and species to which it belongs.
— *Edward Hitchcock*, Ichnology of New England *(1858)*

The first dinosaur trackers

It is probably true to say that the first dinosaur trackers were other dinosaurs. Footprint evidence at several sites indicates that carnivorous dinosaurs sometimes followed herbivorous prey species. Of course, we will never know exactly the hunting or tracking strategies of dinosaurs, whether they used sight, smell, or other senses individually or in combination. However, we can point to a number of probable scenarios based on footprint evidence.

Large and moderate-size carnivorous dinosaurs probably stalked, followed, and hunted in groups or packs. This conclusion is based on a number of sites where theropod dinosaur tracks are found following in the same direction as brontosaur footprints. At the better-known sites in Texas and South America the carnivore tracks overlap the brontosaur footprints, sometimes indenting the raised rims of mud around the herbivore's tracks. Such evidence probably means that the carnivores followed soon after the herbivores.

Roland T. Bird suggested that one of a group of three carnivorous dinosaurs had attacked a brontosaur in a herd of twelve. The footprint evidence from 100-million-year-old Cretaceous coastal plain deposits certainly indicates a group of three predators following a herd of twelve or more brontosaurs, but, as shown in Chapter 13, the evidence for a direct attack on one individual is weak.

The evidence for pack stalking of herds by predators does not imply that individual carnivores were not adept trackers. The Lark Quarry site in Australia (see Chapter 7) reveals the trackway of a supposed single carnivore amid a host of small chicken-size trackmakers that were apparently stampeding away from the threatening theropod.[1] According to the authors of the Australian study, the panic was induced in the small dinosaurs when the large carnivore stalked them on a small lake shore peninsula. If they are correct in their scenario of a predator tracking potential prey into a tight corner, we may infer that dinosaurs had considerable tracking skills, which probably enabled them to survive successfully for so long.

Ancient dinosaur trackers

Early humans were also probably dinosaur trackers, in a sense. Early tribes did *not* hunt these awe-inspiring creatures, of course. Dinosaurs had been extinct for more than sixty million years when our ancestors appeared. Even so, the Holly-wood version of prehistory, with dinosaurs and cavedwellers coexisting, is still one of our most tenacious and outrageous modern myths. However, human trackers must have occa-sionally encountered footprints on rocky pavements and out-crops, although it is doubtful they interpreted them in the way that paleontologists do. It seems likely that these expert trackers of living animals would have been fascinated by fossil tracks of animals they could not place. A number of sites are known where paleolithic rock art exists right next to footprint-bearing outcrops. The most remarkable example, reported from South America by Giuseppe Leonardi, exhibits a circular symbol carved adjacent to a spectacular three-toed carnivore trackway.[2] The well-known resemblance between this type of dinosaur track and those of modern birds, prob-ably led paleo-Americans to believe the trackmaker an exotic giant bird like the living South American Ratite, the Rhea. That was essentially the inference of many nineteenth-century scientists.*

In North America at least two sites in Utah reveal cliff paintings, or petroglyphs, that occur within feet of distinctive dinosaur tracksites.[3] Given the fascination that dinosaur tracks hold for many modern observers, it is probable that early cul-tures also found them intriguing and at least in some cases deliberately chose tracksites as locations for rock art. At a site in Colorado known as Cactus Park (Chapter 4), dinosaur tracks occur within a few yards of a large Ponderosa pine known as the Ute Council tree, a Ute Indian meeting point until com-paratively recent times.

A similar example was reported recently from southern Af-rica. Apparently the Lesotho bushmen have included dino-saur tracks in their cave paintings at sites that are close to well-documented early Mesozoic dinosaur tracksites.[4] In some instances they attempted to reconstruct the animals that had made the tracks and came up with remarkably realistic resto-rations of dinosaurs.

Another interesting example of Paleo-American culture's awareness of dinosaur tracks was reported by Al Look in a booklet on the Hopi Snake Dance.[5] Look illustrated the Hopi dancers wearing "snake priest's aprons" adorned with dino-saur tracks (Fig. 14.1). At present there are several dinosaur tracksites on Hopi Indian land. At least one site near Tuba City is easily accessible to visitors (Appendix A).

In other parts of the world dinosaur tracks have tradition-ally been regarded with awe, reverence, and superstition. For example, in the center of the ancient pueblo of Bretun in northern Spain, Late Jurassic tracks occur in the rocky out-crops. The local inhabitants refer to them as the tracks of an-

*When Scott Madsen was at-tempting to locate a lost site in Ari-zona (see Chapter 4), he learned that Native Americans referred to a site in this area with a name that translates as "location with bird tracks."

Figure 14.1. *Hopi Indians wear aprons adorned with dinosaur track symbols. Artwork by Edward Von Mueller.*

cient pigeons and livestock and believe that they may have been corralled in the town early in its history.[6]

Modern dinosaur trackers

Many modern earth and life scientists, particularly field naturalists, possess the love and awe of nature that probably characterized many ancient tribal cultures. The reaction of the modern dinosaur tracker, stumbling across a new site in the desert, may have been similar to that of our distant paleolithic ancestors. After the initial excitement and curiosity, the objective mind begins to take over. We note, first, the number of tracks, their depth, crispness, variety, and direction. Only then do we think how other members of our scientific tribe might interpret the evidence, a band of modern dinosaur trackers, who, after inspecting such tracks, begin to decipher the evidence, to reason about and record the scientific whys and wherefores for posterity.

The modern scientific approach to dinosaur tracks dates back a little over 150 years. At this time the term *dinosaur* did not exist, and very few dinosaurs were known. Some of the key events and discoveries of ichnology are as follows:

1802. Dinosaur tracks discovered in New England but not studied or documented.

1822. Dr. Mantell discovers the *Iguanodon* in England. It is formally described in 1825.

1836. Edward Hitchcock of Amherst College, Massachusetts, publishes the first description of dinosaur tracks from the Connecticut Valley.[7]

1841. Sir Richard Owen proposes the term *Dinosauria* for classification of *Iguanodon* and a few other forms known by that time.

1846. Iguanodon footprints discovered in England. They are identified as *Iguanodon* tracks in 1862.[8]

1858. First dinosaur, *Hadrosaurus*, discovered in eastern North America. Hitchcock publishes his major work, *The Ichnology of New England.*

1866. First dinosaur tracks reported west of the Mississippi.[9]

1880. First dinosaur tracks reported from Africa.[10]

1882. First dinosaur tracks reported from the Soviet Union.[11]

1899. First tracks reported from the Morrison Formation.[12]

1916. First tracks described from Texas.[13]

1929. Teilhard de Chardin and C. C. Young report the first dinosaur tracks from China.[14]

1936. First dinosaur tracks reported from South America.[15]

1933. First dinosaur tracks reported from Australia.[16]

1935–8. Brontosaur tracks discovered in Colorado and Texas.[17]

1953. Richard Swan Lull publishes his third revision of Hitchcock's monumental work.[18]

1964. Rodolfo Casamiquela publishes a book on modern and fossil tracks[19] and brings new South American track evidence to light.

1971. Hartmut Haubold publishes a major monograph on fossil amphibian and reptile tracks.[20]

1984. Haubold extensively revises and updates his monograph.[21]

1986. The first international symposium on dinosaur tracks is convened, with a week-long field trip through Arizona, Utah, Colorado, Oklahoma, New Mexico, and Texas.[22] 150th anniversary of first paper on subject.

1987. Giuseppe Leonardi and his colleagues publish a comprehensive manual on fossil footprints.[23]

1988–9. Megatracksites reported in western United States.[24]

1989. Dinosaur Tracks and Traces published, the first book devoted entirely to the subject of dinosaur ichnology.[25]

1990. Dinosaur Tracks published, the second book on the subject.[26]

1991. The one hundred fiftieth anniversary of the naming of dinosaurs. *Tracking Dinosaurs* published, the third book in three years on the subject.

There is more to the unfolding of a science than a plain chronology of events. Geographical circumstances and personalities come into play. As indicated, the first dinosaurs were discovered in England, in deposits of Early Cretaceous and Late Jurassic age, whereas the first tracks were discovered in North America in deposits much older, dating back at least 50 million years earlier to the Early Jurassic epoch. One consequence of this disparity was that the connection between dinosaurs and their tracks was not made immediately.

Edward Hitchcock

Edward Hitchcock (1793–1864) was unquestionably the first great modern dinosaur tracker (Fig. 14.2), and his work has been called monumental, a tribute to its extensiveness and thoroughness. Although Hitchcock carefully described thousands of tracks, surprisingly and interestingly he never attributed any to dinosaurs, not even toward the end of his career, when dinosaurs were better known and their tracks were also being discovered and identified in Europe. Instead, he attributed them to birds, not to species closely related to modern varieties, but to forms that might have resembled the extinct New Zealand moa *(Dinornis)* or the elephant bird *(Aepyornis)* of Madagascar (Fig. 14.3). This is not surprising because both these fossil birds were discovered while Hitchcock was working on the New England tracks, and there was much speculation consequently about races of giant birds predating the moa and elephant bird. Most leading scientists of the day supported Hitchcock's interpretations,[27] and with good reason; as we shall see, he was essentially correct.

Tracks were first discovered in New England in 1802 by a farmboy, Pliny Moody. They were small and "turkeylike," causing the neighbors to remark that in order to impress their tracks in stone Mr. Moody must have very heavy poultry! The tracks also came to be known as those of Noah's raven, stressing their birdlike characteristics and ante-diluvian antiquity.

Edward Hitchcock was only a nine-year-old boy in 1802, so it was not until the 1830s that he became passionately interested in footprints and acquired the Moody track slab along with many others. However, almost from the day he became established in the field he became embroiled in a controversy that caused him much consternation and heartache. The problem began when a Dr. James Deane reported these tracks to Hitchcock, suggesting they were turkey tracks worthy of study by a qualified expert. Hitchcock took up the challenge and was duly respectful and appreciative of Deane's contribution in reporting the tracks, even though Deane showed no inclination to engage in serious study at that time. However, Hitchcock knew that they were not turkey tracks even

Figure 14.2. Edward Hitchcock, the father of modern research on Mesozoic fossil footprints.

Figure 14.3. *The elephant bird (left) and moa (right), extinct flightless birds.*

though they were turkeylike, and he spent at least six months studying specimens before drawing any conclusions.

When he published his first account in 1836, he called them *Ornithichnites*, meaning "stony bird tracks." However, until his dying day Hitchcock expressed doubts about the origin of the tracks and wished to make a distinction between true bird tracks and the fossil stony "bird" tracks and make it clear that Jurassic birds were not necessarily like modern turkeys. It was this lack of conviction that later prompted Deane and his friends to claim that it was Deane, not Hitchcock, who had first recognized them for the bird tracks they were. A contentious and sometimes bitter debate ensued, much of it recorded in the pages of scientific publications. In 1858 Hitchcock devoted ten pages to a justification of his position, but the whole dispute was senseless. Even a child could recognize the tracks as turkeylike. Hitchcock's point was that there was then no evidence of Jurassic birds, so the trackmakers could not be interpreted as true birds; at best, they represented bird ancestors. When *Archaeopteryx* was discovered in 1861, two years before his death, Hitchcock claimed his interpretations had been vindicated. Here was proof of a Jurassic bird ancestral to all other forms and, of course, different from modern turkeys and ravens.

As it turns out, Hitchcock's scientific caution was justified. After a spate of late-nineteenth-century dinosaur discoveries, it became abundantly clear that the tracks were dinosaurian in origin, made, in the majority of cases, by bipedal carnivorous dinosaurs. This led initially to the perception that Hitchcock and others of his generation had been mistaken. In the latter part of the twentieth century scientific reasoning has come full circle. It has been convincingly argued that birds are the direct descendants of a group of carnivorous dinosaurs known as coelurosaurs.[28] Because dinosaurs came first, it is scientifically acceptable for modern paleontologists to classify a robin or a seagull as the theropod dinosaur. In one recent scientific paper the authors spoke of eating dinosaur at Thanksgiving and having a dinosaur bath in the backyard.[29] Frivolity aside, the new classifications mean that Hitchcock was right all along: The tracks were indeed those of special types of bird; not creatures identical to modern turkeys, but their ancient theropod ancestors.

During the course of his career Hitchcock amassed a remarkable collection of Early Jurassic tracks; these are still housed at Amherst College, Massachusetts (see Appendix A). The museum, known as the Appleton Cabinet, was designed to Hitchcock's specifications, to ensure that the track-bearing slabs would be illumined by natural light striking them at the optimum angle. Like other dinosaur trackers, Hitchcock had learned that low-angle illumination shows the best detail, so arranged the slabs perpendicular to the windows for the best effect.

Hitchcock's major work, *The Ichnology of the Connecticut Valley*, published in 1858, is considered a classic and is still available in a reprint series.

Sir Arthur Conan Doyle

Sir Arthur Conan Doyle (1859–1930), famous for creating the supersleuth Sherlock Holmes, who possessed, among other talents, superior tracking skills,[30] was also keenly interested in science. In their fascinating study Dana Batory and William Sarjeant show the influence of the true-life scientific exploits of famous naturalists like Charles Darwin on Conan Doyle's work.[31]

Conan Doyle was also influenced by discoveries closer to home, including the unearthing of Cretaceous *Iguanodon* tracks right on his doorstep, near his house in Sussex in 1909. After hearing of the footprint discovery in the nearby quarry, Conan Doyle wrote to Arthur Smith Woodward of the British Museum and persuaded him to visit and inspect the tracks. Conan Doyle later proudly displayed casts of the tracks at his home.

It was the adventures of real-life scientists and these real footprint discoveries that provided much of the inspiration for his *The Lost World*, first published in 1912.[32] In this much acclaimed – and lively – story, Professor Challenger leads an expedition to a remote South American plateau and discovers dinosaurs coexisting with cavedwellers and prehistoric animals; movie makers have had a field day ever since with this highly unauthentic combination. Conan Doyle was much more scientific in his treatment of tracks, however. In one scene the explorers come across a set of adult tracks flanked by the par-

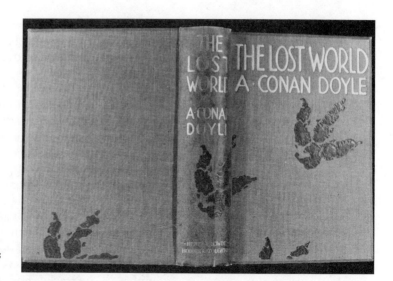

Figure 14.4. *Dinosaur tracks adorn the cover of Sir Arthur Conan Doyle's* The Lost World.

allel trackway of a juvenile. Challenger identifies the small front footprints of the dinosaurs and exclaims that he recognizes the trackmaker from his familiarity with the Wealden footprints (made by Cretaceous *Iguanodon*-like dinosaurs). Given that the co-occurrence of adult and juvenile trackways and iguanodontid front foot impressions were not yet known, or at least not documented in any detail, in 1912, Conan Doyle's creative observations have a mildly prophetic quality. Although some may say that he has no claim to involvement in the serious science of dinosaur tracking, it is worth pointing out that Conan Doyle was the only person ever to record these particular Wealden tracks; he illustrated them in a later edition of *The Lost World* (Fig. 14.4). More important, he made a major contribution to the popularizing of dinosaurs and prehistoric animals and all that is exciting and romantic about their discovery and interpretation.

It is also worth noting that both Conan Doyle and Woodward were involved in another scientific saga at this time, the notorious Piltdown Man hoax. Fragments of this purported ape-like human ancestor were first discovered in 1908 and 1909, again in quarries near Doyle's home in Sussex. Although the fraud was not exposed until 1953, it has become one of the most famous paleontological whodunits. Many people have theories as to who perpetrated the hoax, and Conan Doyle has been implicated on more than one occasion.[33] Again, however, the point is not that Conan Doyle may have been guilty of a massive hoax, but that he was involved in scientific and intellectual endeavor.

Roland T. Bird

The story of Roland Bird's tracking years is one of patience and dedication rewarded. He worked at a time when dinosaur track research was in the doldrums. Although a few paleontologists and geologists had dabbled in dinosaur track research in the latter part of the nineteenth century and the early part of the twentieth century, the science lapsed into obscurity, and little memorable work was published for several generations after Hitchcock's death. Richard Swann Lull revised and updated Hitchcock's work and attributed many of the tracks to various groups of dinosaurs that were becoming known early in the twentieth century, but little new information emerged from the classic sites in New England.

It was in the thinly populated Wild West that exploration began advancing into exciting new areas of dinosaur track discovery. In the 1920s Charles Sternberg reported a number of footprints from Cretaceous deposits in Alberta and British Columbia.[34] The tracks, now known to number in the thousands, included those of large and small theropods, ornithopods, and a probable quadruped.

Although Lull, Sternberg, and others made significant contributions, it was the discoveries and work of Roland T. Bird that are best remembered and probably the most significant. In the 1930s and 1940s Bird worked for Barnum Brown of the American Museum of Natural History as a field collector. Bird was Brown's right-hand man and an invaluable assistant who scoured large tracts of the West in search of dinosaurs and their tracks. It was Brown's intention to have Bird collect representative dinosaur tracks from all over North America and exhibit them in conjunction with the skeletons mounted at the American Museum. To this end Bird traveled widely in the West, pinning down the location of important tracksites in Arizona, Utah, Colorado, Wyoming, and Texas.

Bird's first major footprint project began in 1937.[35] Learning that many large three-toed tracks had been discovered in the Late Cretaceous coal mines of Colorado, Utah, and Wyoming, Brown and Bird decided to excavate in the Red Mountain mine near Cedaredge, Colorado, where a long-stepping trackmaker was reported to have left its footprints. (As discussed in Chapter 13, Brown deliberately dramatized this "Mystery Dinosaur," making a fuss about the giant three-toed tracks in order to attract publicity and funding.) In order to extract these tracks, Bird and Brown required the assistance of skilled miners to remove a 17-foot-long slab containing two good tracks, with all the rock surface in between. This excavation was no small undertaking. They required the help of three round-the-clock shifts of miners and a specially constructed mine cart that was stronger than the standard coal wagon. The slab was eventually extracted in three pieces, with the two exhibit-quality tracks at opposite ends and other, less distinctive footprint features in the middle portion. As with many footprint discoveries, the slab proved controversial, as we have seen (Chapter 13).

Bird, who was both modest and unassuming, published nothing on these tracks at the time, although his posthumous autobiography contains fascinating accounts of the coal mine exploration and the real dangers of extracting tracks and large plant fossils from collapsing mine ceilings.[36] Keeping a characteristic low profile, Bird avoided the trap of exaggerated, unfounded claims, to which Brown succumbed.

Bird's long apprenticeship served him well, and in 1938 he left the Colorado Plateau to follow up a report of tracks in Texas. He had first got wind of these tracks when stopping at the roadside in Gallup, New Mexico, where he observed a magnificent three-toed carnivore track. Little did he know what lay ahead. With characteristic patience and thoroughness, he detoured through southeast Colorado to the Purgatoire site to check out another report of possible brontosaur tracks, before finally arriving at the carnivore trackmaker site at Glen Rose in Texas.

However, the carnivore tracks, situated in the limestones of the bed of the Paluxy River, were not destined to hold his attention for very long. One day while on a lunch break he began to ponder the significance of some large holes or depressions into which he had been shoveling mud and dirt as he uncovered the three-toed footprints. Suddenly it dawned on him: These were tracks, probably those of brontosaurs (Fig. 14.5). Once they were uncovered to reveal the characteristic trackway pattern of small front feet and giant back feet, there was no doubt about the origin of these tracks. Strange as it might seem, tracks of the giant brontosaurs had eluded scientific discovery even a century after those of comparatively small birdlike forms had become widely known. Bird had made the find of his career.

With characteristic dedication and tenacity he stayed up all night on one occasion to make giant plaster casts before the unpredictable Paluxy River rose and flooded. He built a fire for light and companionship, drained oil from his car to grease the tracks before pouring the plaster, and finished the casting project at sunrise.

Bird was a seasoned excavator of dinosaur bones and had made detailed and meticulous maps of bone-strewn quarries. However, documenting dinosaur tracks was a new field with few precedents or guidelines for standardized procedures. In addition, the tracks in the Paluxy River bed were always at least partially submerged. In order to uncover the tracks, map

Figure 14.5. *Dinosaur tracker Roland T. Bird at the Paluxy River site, now Dinosaur State Park, where he made his famous discovery of brontosaur tracks.*

their positions and proceed with their excavation Bird organized a team of Works Progress Administration workers to sandbag the river and clean off the track bearing surface. This allowed recognition of about a dozen parallel brontosaur trackways trending in one direction. A carnosaur trackway converged with a brontosaur trackway and at the point of intersection there appeared to be a footprint missing from the otherwise regular sequence. From this Bird inferred that the carnivore had attacked the brontosaur, and this is how he and Brown decided to extract a portion of the two trackways for an American Museum exhibit, which led to so much debate (see Chapter 13).

The existence of brontosaur tracks at the Texas site suggested that sauropods walked on land, which was contrary to scientific belief at the time. It was widely held that in order to support their titanic weight, brontosaurs were aquatic or semiaquatic swamp wallowers. After examining this new evidence Bird had a hard time envisioning a fully terrestrial brontosaur. In order to explain the lack of tail drag marks he favored compromise, suggesting that the Texas sauropods may have walked in shallow water, floating their tails behind them. The title of his first paper, "Did Brontosaurs Ever Walk on Land?"[37] clearly suggests an intellectual dilemma, and his discovery had revolutionary implications for the science.

In addition to being the first to recognize brontosaur tracks and use them to dispute the aquatic sauropod hypothesis, Bird also made another major contribution to dinosaur paleoecology. At another Texas site, Davenport Ranch, he uncovered twenty-three sauropod trackways heading in the same direction. Bird rightly concluded that such evidence strongly suggested herding behavior. Again, such a concept was revolutionary in the 1930s; Bird was characteristically cautious about discussing the implications of gregariousness among dinosaurs. We now know that dinosaur gregariousness was quite common (see Chapter 7), and we recognize Bird as one of the originators of the herding hypothesis and as the first paleontologist to present largely convincing evidence.[38]

Late twentieth-century dinosaur trackers

Until recently there was no book on dinosaur tracks and few on fossil animal tracks, even though an extensive literature describes a variety of often inconspicuous traces and trails of invertebrates like worms and crustaceans. This emphasis began changing in the past few decades due to the contributions of a diverse group of dedicated trackers. For want of a better term we shall call this mixed bag of detectives the seven sleuths. In alphabetical order they are Dr. Donald (Don) Baird, recently retired professor emeritus at Princeton; Rodolfo Casamiquela, pioneer South American tracker and author of *Estudious ichnologicos;* Dr. George Demathieu, one of France's most prolific and substantive ichnological authors; Dr. Paul Ellenberger, another prolific French author who took it upon himself to document the extensive track-bearing layers of Southern Africa; Dr. Hartmut Haubold from East Germany, author of *Ichnia Amphibiorum et Reptiliorum fossilum* (Fossil tracks of amphibians and reptiles) and *Saurierfahrten* (Reptile tracks), the first comprehensive compilations on fossil amphibian and reptile tracks; Father (Dr.) Giuseppe Leonardi, South America's leading ichnologist and editor/compiler of the *Glossary and Manual of Tetrapod Footprint Paleoichnology;* and finally, Dr. William (Bill) Sarjeant, expert on British tracks and the obscure history of early research on fossil footprints.

This heterogeneous international group has worked individually and collectively over the past generation to promote some degree of consistency and standardization in the science of tracking. The seven have described tracks from all epochs in the Age of Dinosaurs, and other periods besides, and between the 1950s and 1980s they published their findings in hundreds of papers in half a dozen languages. All have made substantial, often innovative contributions to a science that, before their participation, lacked much credibility.

Although it is hard to single out any individual from this group, Giuseppe Leonardi has made a monumental contribution. His father was a tracker in his own right, responsible for documenting Permian tracks from northern Italy. Following in his father's footsteps, and the example of the great Teilhard de Chardin, Giuseppe became a qualified paleontologist (and learned ichnologist) while also studying to become a priest. In the early 1970s he traveled to Brazil with an expedition from the University of Rome in search of meteorites; he later returned to take up citizenship and follow the calling of the Church. Although he had been prepared to give up footprint research, he soon learned that there was no shortage of tracks in that part of the New World. As he began to document his findings and incorporate tracking into his ministerial itineraries, he discovered a variety of dinosaur tracksites from different epochs, as well as a number of other significant footprints, including what may be one of the world's oldest, amphibian tracks from the Late Devonian.

More than most other dinosaur trackers of his generation, Father Leonardi has experienced the rigors, both physical and psychological, of frontier paleontology. Far from the comforts of a well-furnished office he has traveled to the most remote terrain in South America, risking piranha-infested waters and flooding rivers that threatened to cut him off from the only return route to civilization. On one occasion he and his companions were robbed of everything by bandits. They counted themselves lucky to escape with their lives.

Giuseppe Leonardi's work is distinguished by its breadth and innovation. He has described tracks and traces from the Devonian to Recent times. He has been a field pioneer, mapping remote sites in inaccessible jungle locations. Among these are the footprint-rich Carir Basin sites, where some of the first multiple track-bearing layers were accurately documented (Chapter 8), and the spectacular Toro Toro site in Bolivia, where a large herd of theropods were following a group of brontosaurs. In addition, Father Leonardi was one of the first to compile and synthesize footprint data on a continental scale and to recognize various paleoecological and paleoenvironmental trends. As editor of the *Glossary and Manual of Tetrapod*

Footprint Paleoichnology, he was also responsible for standard-izing our technical vocabulary in eight languages, most of which he speaks fluently.

Conclusions

The dedication and resolve of trackers like Hitchcock, Bird, and Leonardi is in no small measure responsible for inspiring new generations to take up the calling. There has probably never been a better decade than the 1990s to respond to this call. The 1980s were a decade of unprecedented advance and progress. In addition to the seven sleuths, most of whom are still actively researching, at least as many could fairly be iden-tified as notable scientists in the current generation, many of whom have been mentioned throughout the text. Suffice it to say that researchers like Phil Currie, James Farlow, George Gand, Shinobou Ishigaki, Joaquin Moratalla, Paul Olsen, Kevin Padian, Jeff Pittman, José Luis Sanz, and Tony Thulborn are among those actively publishing in the 1980s and 1990s.

This brings us up to date on the most important contribu-tors to the science of dinosaur tracking through the twentieth century. On the eve of the new millenium the future of di-nosaur tracking looks bright. We will end by looking at what the future may have in store for dinosaur trackers in the twenty-first century.

15

Epilogue: Trail to the twenty-first century

An erect walking phytosaur, based on Late Triassic footprint evidence. Artwork by Doug Henderson.

These are heady times for those of us with an interest in and love for dinosaurs.
— Dan Chure, in The Age of Dinosaurs *(1989)*

Having, I hope, given credit where credit is due, I will attempt a little prophecy about the areas in which this science will advance in the present decade, taking up the trail to the twenty-first century. Progress will continue to be made in the following areas:

Identification of trackmakers
Standardization and quantification of data (locomotion, census, and so on)
Studies of dinosaur population dynamics and growth rates
Documentation of Mesozoic bird and mammal tracks
Understanding of track preservation
Documentation of new sites, especially in neglected areas
Understanding of megatracksites
Implications of dinosaur tracking for the broader field of earth science
Conservation, preservation, and interpretation

I have listed these areas of interest in much the same order as in the book, from paleobiological to paleoenvironmental. A few words on each of these topics will suffice.

Identification of Trackmakers

As more footprints are found, documented, and identified, we will, by a process of elimination, learn to distinguish carnivore from herbivore, brontosaur from ankylosaur, pterosaur from crocodile. Study of modern tracks and fossil tracks from before and after the Age of Dinosaurs will help us hone our tracking skills. This progress, already underway, will allow us to tackle the thorny problem of naming tracks scientifically touched on in Chapter 5. With luck and perseverance, we should minimize, if not eliminate, inappropriate names and footprint designations or identifications. We will then be able to speak confidently about tracks belonging to a particular family or group of dinosaurs. For example, in the future when we talk about tyrannosaur tracks, voices of dissent will not be raised, because the criteria for their identification will have been established and adhered to with reasonable rigor and certainty. Improved track identification will rely to some degree on progress in some of the following areas of research.

Standardization and quantification of data

Greater standardization and consensus in the field is already evident. It goes hand in hand with increased standardization in measurement and quantification of the available data. Most life scientists have employed statistics and mathematics to help with description, whether of living or of ancient populations, and dinosaur trackers are also beginning to use quantitative analysis to help define footprint and trackway shapes, angles, and other measurements, leading to more reliable syntheses and in-depth understanding. Numbers can also be applied to census counts of the various groups recognized at tracksites. We can talk with increasing confidence of the percentages of different dinosaur groups represented.

Studies of dinosaur population dynamics

Sufficient data exists for many sites allowing trackers studying the range of size of footprints to estimate how many babies, subadults, and adults were active in a particular area. This type of analysis sheds light on the growth, or ontogeny, of individuals and allows researchers to investigate the population dynamics or population structure of dinosaur species. Such studies tell us whether populations were dominated by young individuals, old individuals, or a broad cross section. When combined with data from bone sites and nest sites, the results may shed new light on dinosaur demography, or how dinosaur populations were structured.

Documentation of Mesozoic bird and mammal tracks

It is not just the study of dinosaur tracks that is enjoying a scientific renaissance. Other fossil footprints are also being found and studied seriously. As discussed in Chapters 8 and 9, we have discovered that Mesozoic bird tracks are quite plentiful. In fact, they are so widespread and abundant in Early Cretaceous sites that they tell us much more than bones about evolution. This implies a likelihood that footprints will shed light on the origin of birds and particularly on the radiation and diversification of shore-dwelling varieties.

Since we already know that mammals evolved early in the Mesozoic, we might expect future footprint discoveries to include hitherto unknown mammal tracks. Until now, true Mesozoic mammal tracks have proved highly elusive, but this might simply be due to their small size. Once trackers find out where to look, mammal tracks may prove to be more common than previously supposed.

Understanding of track preservation

Track preservation has been a neglected area, but thankfully we note that is being given more serious attention. As we improve our understanding of how tracks were made and preserved, we will avoid the pitfalls of careless speculation and interpretations that are not borne out by the rock record. In particular we can hope to better understand tracks and undertracks and more fully appreciate the significance and potential of both for a sound scientific interpretation.

Documentation of new sites

The remarkable increase in the documentation of sites continues apace. The discovery of sites in North and South America by trackers like Currie, Leonardi, Pittman, Padian, Olsen, and the University of Colorado at Denver group underscore the fact that tracksites are extremely common and widespread. As yet only a modest amount of information has emerged from certain other areas, however. Whereas sites are abundant and accessible in Europe, North America, and parts of southern Africa, other parts of Africa and Asia, particularly China and the USSR, reveal a rather scanty track record relative to the vast land areas involved. This is almost certainly a result of less intensive study than of a scarcity of sites. We already know of about thirty sites in China, a handful in the USSR, and a growing number in northern Africa. Many of these sites are large and yield thousands of tracks. As they become better known and documented, we can expect to glean valuable data and information suitable for comparison with that available for Europe, the Americas, and elsewhere. Asia has the potential for yielding hundreds, even thousands, of sites, rather than a few dozen.

Understanding of megatracksites

It is also highly unlikely that megatracksites exist in just three regions of North America. As mentioned in Chapter 12, there may be one in North Africa. Moreover if, megatracksites are related to sea-level changes, their location should be fairly easy to predict. Once located, they can be studied in the context of the ancient habitats they represent. Such study should produce large amounts of dinosaur footprint data and also be applicable to sedimentology, stratigraphy, and sea-level research, areas of geoscience that often have a direct bearing on the exploration for fossil fuels and other natural resources.

Implications for earth science

We can also speculate on the usefulness of the new dinosaur footprint data in the broader fields of earth and life science. It is likely that these data will be subjected to various syntheses and analyses that seek to test existing hypotheses about dinosaurs, their behavior, and distribution. A good example would be the recent discovery of tracks right at the Cretaceous–Tertiary boundary, only 37 centimeters, about 1 foot,

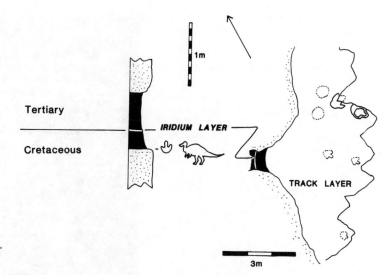

Figure 15.1. *Dinosaur tracks occur only 37 centimeters below the Cretaceous–Tertiary boundary in southeastern Colorado.*

below the famous iridium layer (Fig. 15.1). This layer, caused by fall out from a major catastrophic event, perhaps impact with a meteorite or a volcanic eruption, marks the end of the Mesozoic Era, the end of the Age of Dinosaurs.[1] For over a decade earth scientists have argued about whether the dinosaurs died out gradually before the end of the era or suddenly at the time of the purported catastrophe. Since there are no complete or partial skeletons of dinosaurs in strata for at least three meters below the iridium layer, paleontologists speak of a three-meter gap.[2] However, the discovery of tracks 37 centimeters below the iridium has narrowed the gap to only one tenth of its previous magnitude. The tracks are indisputable proof of live dinosaurs, in this case duckbills, living later in the Cretaceous than was previously proven. A simple discovery of a few tracks (Fig. 15.2) improves our resolution of dinosaur data at the Cretaceous–Tertiary boundary by an order of magnitude.[3]

Recent work has also shown that we have paid insufficient attention to the usefulness of dinosaurs and their tracks in understanding Mesozoic chronology. The science of biostratigraphy uses fossils to help discriminate subdivisions of geologic time. Traditionally dinosaurs have not played much of a role in this field because they are thought of as characterizing long periods of geologic time, such as the Jurassic, Cretaceous, or even the Age of Dinosaurs. Indeed, ignorance of the true age of dinosaurs and their tracks has been a stumbling block for scientists and laypersons alike, as many examples in this book show. As discussed in Chapter 1, long geologic time periods are divided into shorter epochs and ages. If we study the unfolding of evolution and dinosaur commu-

Figure 15.2. Detail of duckbill dinosaur track 37 centimeters below the
Cretaceous–Tertiary boundary.

nity development age by age, we recognize that various dinosaur species are associated with particular stratigraphic
levels. They are what biostratigraphers call index fossils, which
are characteristic of specific "biochrons," that is, ages or time
spans.[4]

The same chronological (or biostratigraphic) principles that
apply to dinosaur skeletal fossils also apply to tracks and trace
fossils. As shown in Figure 15.3, many of the dinosaur tracks
discussed in this book are associated with deposits of a particular age. For example, based on the sequence of strata in
the western United States, Late Triassic deposits mainly contain nondinosaurian tracks dominated by *Brachychirotherium*
(see Chapter 8). Then comes an Early Jurassic age of theropod
tracks, with abundant *Grallator* and *Eubrontes* footprints, followed by levels with *Otozoum* and *Brasilichnium*. In later Jurassic times we see evidence of the rise of sauropod faunas,
and these persist into Early Cretaceous times. Then comes a
succession of ornithopod track-bearing layers; *Iguanodon*- like

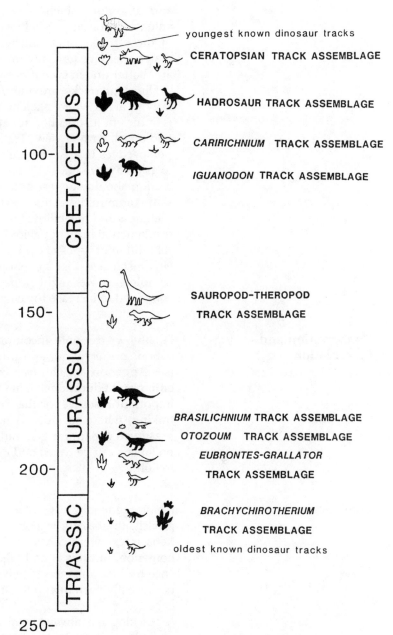

youngest known dinosaur tracks

CERATOPSIAN TRACK ASSEMBLAGE

HADROSAUR TRACK ASSEMBLAGE

CARIRICHNIUM TRACK ASSEMBLAGE

IGUANODON TRACK ASSEMBLAGE

SAUROPOD-THEROPOD
TRACK ASSEMBLAGE

BRASILICHNIUM TRACK ASSEMBLAGE

OTOZOUM TRACK ASSEMBLAGE

EUBRONTES-GRALLATOR
TRACK ASSEMBLAGE

BRACHYCHIROTHERIUM
TRACK ASSEMBLAGE

oldest known dinosaur tracks

CRETACEOUS

JURASSIC

TRIASSIC

100-
150-
200-
250-

Figure 15.3. *Tracking dinosaur communities through the Age of Dinosaurs. Distinctive assemblages of dinosaur tracks help us understand Mesozoic chronology.*

tracks are succeeded by *Caririchnium* footprints and then by Late Cretaceous assemblages of duckbill dinosaur tracks. The youngest track-bearing units are rich in ceratopsian tracks, attesting to one of the last dinosaur communities. In short, we have a sequence of track-bearing layers spanning the Age of Dinosaurs and shedding light on the timing of dinosaur activity epoch by epoch and age by age. As discussed, such data are crucial for research into ancient community ecology and dinosaur habitats, and they are also proving to have valuable applications in biostratigraphy. For example, many of these track zones can be recognized outside the United States, helping us compare chronologies around the world and better understand dinosaur community evolution.

Dinosaur tracks provide much new information, particularly in areas where skeletal remains are lacking. This alone is sufficient justification for studying tracks, but it is a rather traditional, narrow view. Tracks are also providing radical new perspectives in areas where bones are known. We are forced to ask why two lines of dinosaur evidence from the same formation should suggest different conclusions about the dinosaur communities. This question, which pertains to the completeness and reliability of the fossil record, is one of the most fundamental in all of paleontology. The answer appears to be that different types of evidence are preserved under different circumstances, that no particular type of evidence can be complete as we might hope. All lines of evidence must be combined to increase the completeness of the picture.

Conservation and Preservation

Finally, we can note encouraging trends towards the conservation, preservation, and interpretation of many track sites (see Appendix A). This trend results from recognizing the scientific and educational value of tracks and helps preserve an important resource for the future. Particularly in the field of traking, where so much is new, important sites need to be preserved for future generations. In this way current interpretations can be tested and evaluated as new evidence comes to light.

Conclusion

I have generally advocated an holistic and synergistic approach, the broad, or global, perspective. For many generations tracks were considered an obscure and separate phenomenon, but this is no longer the case. In an era when science has become increasingly interdisciplinary, track studies have finally been integrated into the mainstream of earth science.

Tracking has always been an important human endeavor or skill employed by hunters and modern detectives in solving fundamental problems and ensuring survival. But tracking, both modern and ancient, are threatened by the spread of concrete and tarmac at our feet. The earth is no longer receptive to our footfalls, and we are no longer receptive to the textures of the earth beneath our feet. We are a generation that could walk this earth without leaving a single footprint. Only when we walk beyond the tarmac and remove those shoes that advertise our favorite sports manufacturers will we truly have our feet on the ground and leave our own individual mark. Walk barefoot through a swamp or muddy tidal flat and you may leave your tracks in the timeless geologic record, more indelibly than running a hundred marathons down Main Street.

Our tracking skills are no longer developed by the life-or-death necessity of stalking and hunting, but they are only dormant, not lost. At a time when many modern animals are threatened with extinction, the discovery of new dinosaur species has accelerated at an unprecedented rate. As we block out footprints from our urban and suburban habitats, dinosaur trackers return from the field to report hundreds, thousands even millions of newly discovered footprints. Unlike the blank asphalt pages we produce with each new car park, the dinosaur track-bearing layers are rich tapestries of ancient life, pages filled with clues and stories of bygone activity. Nature is not easily suppressed. If we reduce the diversity of living forms and eradicate track-bearing soils until we reach bedrock, we will inevitably run across the abundant traces of ancient life.

This has been the saga of dinosaur tracking, especially in recent years. Once neglected as an insignificant scientific discipline, renewed interest in dinosaurs has prompted a revolution in tracking and the blossoming and maturation of a branch of earth and life science that is as sacred (entitled to reverence and respect) now as when paleolithic cultures first began carving track petroglyphs. Our ancestors relied on their tracking skills and knowledge for survival and gained thereby a deeper understanding and reverence of the natural world. We find ourselves in a similar position with fossil tracks. They are more than mere curiosities and demand more than frivolous and childish interpretations. To satisfy our curiosity about dinosaurs we hone our tracking skills, we hope in the same spirit as our ancestors. As we do so our understanding matures and the rich tapestry of tracks is read with a fuller and deeper appreciation.

"There is no branch of detective science which is so important and so much neglected as the art of tracing footsteps"
Conan Doyle 1891

Appendix A: Where to visit dinosaur tracksites

By no means have all dinosaur tracks been reported, studied, or collected for museums or other institutions. Given the abundance of fossil tracks worldwide and the spectacular rate of discovery in recent years,[1] discovering new sites is quite possible for alert, observant enthusiasts who know where and how to look. However, it must pointed out that finding tracksites is not easy, even for experienced geologists. For amateurs who do not have the time or inclination for much field exploration, there are many easily accessible, well-known museums, parks, and field sites. The following list, including field and museum exhibits scattered around the world, is by no means exhaustive. All sites have been selected because they include some form of professional interpretation.

**Alameda Parkway
(Dinosaur Ridge)**

Situated just west of Denver near the site of the famous 1877 discovery of Jurassic dinosaur bones in the Morrison Formation, the younger Cretaceous Dakota Group beds of Alameda Parkway exhibit several hundred footprints of iguanodontid and coelurosaurian dinosaurs. The site is fenced off, and interpretative brochures and a "Field Guide to Dinosaur Ridge"[2] are available. As the guide explains, the Alameda ("promenade") site really was a dinosaur promenade 100 million years ago. It is part of a megatracksite that has been dubbed the Dinosaur Freeway.

**The American Museum of
Natural History**

Although not best known for its fossil footprints, the American Museum of Natural History in New York boasts a spectacular dinosaur tracks exhibit prepared by Roland T. Bird, showing a Jurassic *Diplodocus* mounted above Cretaceous brontosaur tracks from Texas. As discussed in Chapter 13, the exhibit is unauthentic for a number of reasons; however it does include the best brontosaur trackway segment on display anywhere in the world.

The Appleton Cabinet

A remarkable monument to the great ichnologist Edward Hitchcock (Chapter 14), the Appleton Cabinet in Amherst, Massachusetts, is unique, a museum built around a large collection of fossil footprints. The track collection is probably the world's largest and is undoubtedly one of the best-documented, most famous, and historically most important. The tracks are for the most part those of Lower Jurassic dinosaurs (*Anchisauripus*, *Eubrontes* and *Grallator*), but there are also some made by other contemporaneous animals from the Connecticut Valley.

Barkhausen

Situated near Osnabruck in Germany, this locality features an on-site exhibit of Late Jurassic sauropod footprints and three-toed carnivore tracks. The tracksite exhibits one of the few examples of sauropod tracks in Europe.[3] A booklet and interpretative materials are available.

Dinosaur State Park, Connecticut

An on-site exhibit of Lower Jurassic tracks similar to those studied by Edward Hitchcock can be found at Rocky Hill, Connecticut. As explained in Chapter 4, the trackbearing layers were uncovered by highway construction activities in 1966, and authorities moved with admirable dispatch to preserve the site as Dinosaur State Park and interpret it for the benefit of the public.[4] Dinosaur tracks at this site include *Anchisauripus*, *Eubrontes*, and *Grallator*.

Dinosaur State Park, Texas[5]

Situated on the banks of the Paluxy River near Glen Rose, this is the tracksite made famous by Roland T. Bird when he first uncovered and described spectacular Cretaceous brontosaur tracks, some of which were excavated and taken to the American Museum. Much of the track-bearing layer is under water and river valley alluvium, but replicas are on display at the park entrance, and rangers are available to advise visitors as to what can be observed.

Dinosaur Valley

Named for the dinosaur-rich Grand Valley, this branch of the Museum of Western Colorado in Grand Junction contains a diverse collection of dinosaur tracks and a modest dinosaur footprint exhibit. The museum is close to several well-known dinosaur localities, so visitors can walk along a number of outdoor dinosaur trails.

Dorchester Museum

The Dorchester Museum, in Dorset County, is small, but its modest exhibit is representative of the dinosaur tracks found along the south coast of England. Miscellaneous footprints and fossils are also on display in a number of small towns, from Dorchester and Lyme Regis in the west, through the Isle of Wight region, and east as far as Sussex.

Dr. Alf's Museum

Situated in Claremont, California, Dr. Alf's Museum, a part of Dr. Alf's School, is a remarkable institution, a monument to the energy and imagination of the school's founder, the indefatigable Dr. Alf. The museum contains a large and diverse collection of fossil tracks, including many that are pre- and post-Mesozoic. Many of the tracks were collected by Dr. Alf and his students, and some of the fossil mammal tracks have been described by Dr. Alf in the scientific literature. The best tours have always been those introduced or conducted by Dr. Alf himself.

Lark Quarry

The Middle Cretaceous Lark Quarry dinosaur tracksite is situated in a remote region of Queensland, Australia. Despite this fact, it has been extensively documented,[6] and considerable efforts have been made to protect and preserve the site. It is famous for the large number of small dinosaur tracks that have been interpreted as evidence of a dinosaur stampede (see Chapter 7). Explanatory signs have been posted at the site, and replicas of tracks are on sale at the Queensland Museum in Brisbane.

Moab

Dinosaur Tracks are abundant in the vicinity of Moab, Utah. It is one of several areas in the Colorado Plateau Region that bristles with track sites. In the Moab area, sites range in age from Late Triassic to Cretaceous.[7] Currently the famous Jurassic-age purported pterosaur tracks are on display outside the Bureau of Land Management office (see Chapter 13). Another Jurassic site, about 7 miles south of the town, is marked with a road sign, and visitors can inspect a variety of carnivore tracks at first hand. Visitors can also visit the "turning brontosaur" tracksite situated about 20 miles north of town (see Fig. 6.4).

Münchehagen

Situated near Hannover, Germany, the Münchehagen site features a spectacular exposure about the size of a football field, with seven sauropod trackways and a carnivore trackway. One of the sauropod trackways is protected by a permanent shelter. The site is Lower Cretaceous in age. Brochures are available at the site, and the tracks are documented in detail in the German literature.[8]

Ribadessella

Recently the Spanish Institute of Geology and Mineralogy published an impressive color booklet on the dinosaur footprints of the Ribadessella area in northern Spain,[9] describing, among other things, the variety and significance of Late Jurassic herbivore and carnivore trackways exposed in the spectacular cliffs and coastal exposures. The area is designated a point of geological interest.

Price

The College of Eastern Utah Prehistoric Museum is situated in Price, Utah, at the center of the Carbon County coal mining district. Generations of miners have collected Late Cretaceous tracks from many coal seams. Visitors will find over 50 tracks on display in the newly expanded exhibit. The museum is part of the so-called Dinosaur Triangle between Dinosaur Valley (Colorado) and Dinosaur National Monument near Vernal, Utah.

La Rioja

The somewhat younger (Early Cretaceous) group of tracksites and found in the La Rioja region of north-central Spain has been described in a number of technical publications as well as in an attractive popular guidebook.[10] Like the Ribadessella sites, the La Roija sites yield a variety of herbivore and carnivore trackway types. They are also accessible to the public and marked and protected with signs, fences, and shelters.

Samchampo

Exposures of the Cretaceous Jindong Formation near Samchampo on the southern coast of South Korea are replete with dinosaur track-bearing layers (see Chapter 10). The area is easily accessible and is frequented by many Korean tourists. Recently a permanent explanatory sign (in Korean and English) was erected next to the footprint-bearing layers. The site is extensive and reveals tracks of various herbivorous dinosaurs (sauropods and ornithopods), a few carnivorous species, and birds.[11]

Tuba City

Situated on Navajo tribal lands just west of the small town of Tuba City, Arizona, a number of Lower Jurassic dinosaur tracksites are open to public viewing. Tracks at the most frequently visited site are easily accessible and marked by roadside signs. The tracks have been studied and named as *Kayentapus* (from the Kayenta Formation) and *Dilophosauripus* (*Dilophosaurus* tracks).[12]

Tyrrell Museum of Paleontology

The Tyrrell Museum of Paleontology in Alberta, Canada, contains a large collection of dinosaur footprints, including some on display. The collection is particularly important because of the large number of specimens that were collected from the Peace River,[13] British Columbia, before the valley was flooded in the 1970s. The museum also contains the largest dinosaur exhibit in the world.

Vernal

The Field House in Vernal, Utah is a Natural History Museum at the center of northeastern Utah's "Dinosaur Land." The exhibit includes several dozen dinosaur tracks ranging in age from Triassic to Cretaceous. The museum is close to the famous quarry at Dinosaur National Monument and is part of the aforementioned Dinosaur Triangle.

Appendix B: Glossary

Terms used in describing fossil footprints and tracksites

Bed:
A single layer of sedimentary rock or a single stratum; it can vary greatly in thickness and composition. As with most sedimentary layers, the original sediment was deposited on a horizontal plane in the majority of cases.

Bedding plane:
The planar surface that exists between two beds, or layers of strata. Typically exposed by erosion at the earth's surface, bedding planes often represent former substrates, or surfaces on which animals left their footprints.

Biostratigraphy:
The dating and zoning of strata using fossils.

Bioturbation:
Disturbance of soils and substrates by animals (and plants); describes invertebrate burrowing as well as tracks and traces of vertebrates.

Biped:
An animal that walks on its two hind feet. *Bipedal* describes such an animal.

Body fossil:
Part of the actual remains of an animal, usually bones and teeth, but sometimes mummified skin or tissue.

Cast:
The negative replica of a fossil impression or mold. The lithified infilling of a footprint is a natural cast.

Coprolite:
Fossil dung, feces, or spoor, not necessarily found with tracks.

Deposit:
Generic term for a sedimentary deposit (e.g., sandstone, mudstone, or limestone), laid down or deposited in a particular area whether of small or large extent.

Digit:
Individual finger or toe.

Digitigrade:
Walking on the digits only

Dinoturbation:
Trampling and disturbance of soils and substrates by dinosaurs.

Facies:
Term denoting sedimentary deposit of a particular character and or genesis, for example, deltaic facies or playa lake facies. Cf. Deposit.

Footprint:
The impression made by any foot (front or back) in soft sediment. It subsequently becomes a fossil footprint when the sediment turns to rock. May be used loosely to describe a footprint cast. (Synonym: track.)

Formation:
A formal unit of geological stratum that is distinct from other units and widely recognized; the formal designation of a sedimentary deposit, usually named after a location where the formation is well represented, for example, Morrison Formation.

Fossil footprint:
See Footprint; Track.

Gait:
Type of progression, for example, walking, trotting, running.

Ghostprint:
See Underprint.

Hand:
Informal term to describe the front foot, or manus, in tetrapods. Used most frequently to designate front foot of a bipedal animal.

Hop:
A bounding gait, where an animal pushes off with both hind feet at once for a short airborne phase before landing simultaneously on both feet.

Ichnology:
The study of trace fossils, including tracks and other traces.

Manus:
Latin: "hand." Refers to front feet in four-footed animals; not necessarily used for walking.

Megatracksite:
A very large or regionally extensive tracksite associated with a particular layer of strata. May be exposed at the surface as a series of smaller tracksites separated by erosion and topography.

Overprint:
When an animal partially or completely overprints or overlaps its own front footprints with its hind footprint.

Pace:
A single step, left foot to right or right to left; can also imply speed, as in fast or slow pace progression.

Pace angulation:
The angle made by steps relative to the direction of progression. Angle made in the V drawn between a left, a right, and another left footprint (L-R-L) or a right, a left, and another right footprint (R-L-R).

Pes:
Latin: "foot." Refers to hind feed in four-footed animals.

Plantigrade:
Walking with digits flat and part or all of the ankle and wrist also on the ground.

Progression:
Forward movement in a particular direction.

Quadruped:
An animal that walks on all fours. *Quadrupedal* describes such an animal.

Run
Fast progression in which an animal's feet leave the ground for short phases in each step.

Speed:
Measure of the absolute rate of motion of an animal relative to a stationary point, measured in miles per hour or kilometers per hour.

Spoor:
The feces, tracks, and trails left by modern animals, not usually used in discussion of fossil tracks.

Step:
Motion of putting one foot in front of another, or more commonly a measure of the distance from left to right footprints (L-R).

Strata: Collective term for layers, or beds, of sedimentary rock.

Stratigraphy: (From *strata*) The study of stratified sedimentary rock sequences.

Stride: Two steps right-left-right (R-L-R) or left-right-left (L-R-L), that is, the distance between one footprint and the next footprint made by that same foot.

Substrate: The ground on which animals walk or progress; can be soft (incompetent) or firm (competent), on land and even underwater (as a lake- or a seabed).

Tetrapod: A four-legged or four-limbed vertebrate that may use all legs for walking (quadruped) but may also be bipedal or show other adaptations.

Trace: A vague term denoting marks or spoors left by animals; includes tracks and all other marks.

Track: The impression made by any foot (front or back) in soft sediment. Subsequently becomes a fossil track when the sediment turns to rock. (Synonym: footprint.)

Tracksite: A location where tracks are found; normally denotes a fossil tracksite.

Trackway: Two or more consecutive footprints (steps) belonging to a particular animal progressing in a given direction.

Trail: Sometimes used to denote a trackway, but also indicating the continuous marks made by invertebrates (e.g., worms or snails) in which no footprints exist. Also used to denote a well-worn path, for example, a game trail.

Underprint: The impression made by a footprint on an underlying layer of strata. May closely resemble a true footprint or be quite indistinct. (Synonym: ghostprint.)

Walk: Slow progression in which the leading foot (or feet) touches the ground before the trailing foot (or feet) lifts off.

Terms used in describing main dinosaur groups

Ankylosaur: Armored quadrupedal ornithischian with ankylosed, or fused, plates.

Brontosaur: Large saurischian, for example, *Brontosaurus*. See Sauropod.

Carnosaur: A robust type of carnivorous theropod.

Ceratopsian: Horned Cretaceous quadrupedal ornithischian.

Coelurosaur: Gracile, hollow-boned theropod.

Dinosaur: Member of Dinosauria, a group of archosaurian reptiles that are morphologically distinct from other major reptile groups, such as lizards and snakes. Surviving archosaurs include the crocodiles and alligators. Does not include pterosaurs, turtles, extinct marine reptiles, or mammal-like reptiles but, according to some authorities, can include birds.

Hadrosaur: Usually refers to Late Cretaceous duckbill dinosaurs; belongs within Ornithopoda; probably descended from the iguanodontids.

Ornithischian: Bird-hipped dinosaur.

Ornithopod: Means "bird foot"; includes bipedal ornithischians like the iguanodontids and hadrosaurs.

Saurischian: Reptile-or lizard-hipped dinosaur.

Sauropod: Means "reptile foot"; refers to Jurassic and Cretaceous long-necked, long-tailed herbivorous saurischians, also called brontosaurs.

Theropod: Bipedal, carnivorous saurischians; includes gracile coelurosaurs and robust carnosaurs.

Notes

Chapter 1

1. Only a single dinosaur bone had been discovered before 1802. The significance of this early discovery was unknown until much later.
2. A glossary of terms is presented as Appendix B.
3. Traditionally zoologists and paleontologists classify vertebrates into major groups, e.g., Class Amphibia, Reptilia, Aves, or Mammalia. In a controversial paper, Robert Bakker and Peter Galton, proposed the Class Dinosauria encompassing dinosaurs and birds. Many paleontologists feel this inflates the importance of dinosaurs relative to other classes (e.g. Reptilia). Subclass Dinosauria is probably quite sufficient in most people's classification. Bakker, R. T., and Galton, P. 1974. "Dinosaur Monophyly and a New Class of Vertebrates." *Nature* v. 248 p. 168–172.
4. Quoted from Halfpenny, J. 1986. *A Field Guide to Mammal Tracks in Western America.* Boulder, Colo., Johnson 161 p.
5. One of the best dinosaur texts is Norman, D. 1985. *Dinosaurs: An Illustrated Encyclopedia.* New York: Crescent Books.
6. Technically sedimentary rock with fossil tracks can be metamorphosed into metamorphic rock, but there are few documented examples of such metamorphism. The author has also observed dinosaur tracks in igneous rock, in a Cretaceous age sill in the Jindong formation of South Korea. Here, molten magma intruded between layers of sedimentary rock, producing a cast of the underside of a brontosaur trackway. This can be seen as impressions on top of the igneous sill rock.
7. Philip Currie, assistant director of Alberta's well-known Tyrrell Museum of Paleontology has reported examples of dinosaur tracks that were eroded out of Cretaceous strata and redeposited in young alluvial deposits of Pleistocene age. Currie, P. J. 1989. "Dinosaur Footprints of Western Canada." In Gillette, D. D. and Lockley, M. G. (Eds.) *Dinosaur Tracks and Traces.* New York: Cambridge University Press, pp. 293–300. As discussed in Chapter 15, if tracks are found above the Cretaceous–Tertiary boundary, they will prove once and for all that dinosaurs survived the terminal Mesozoic catastrophe.
8. Geologists and paleontologists date rock strata using a combination of tried and tested methods, including radiometric dating and biostratigraphy.
9. One of the best examples was presented by Paul Olsen and Peter Galton. They compared Triassic and Jurassic tracks from New England and Southern Africa, establishing and clarifying the age of strata in both regions. Olsen, P., and Galton, P. 1984. "Review of the Reptile and Amphibian Assemblage from the Stormberg of South Africa, with Special Emphasis on the Footprints and the Age of the Stormberg." *Paleontologica Africana* v. 25, p. 87–110. See also Padian, K. 1986. *The Beginning of the Age of Dinosaurs.* New York: Cambridge University Press, 378 pp.
10. For many years "mammal-like-reptile" has been used to describe the ancestors and relatives of the true mammals that lived from Permian times through to the middle of the Age of Dinosaurs. Some say the are really reptile-like mammals. Here, I simply use the term "mammal-like." Padian, K., and Chure, D. J. (Eds.) 1989. *The Age of Dinosaurs.* Short Course no. 2. Knoxville, Tenn.: Paleontological Society, 210 pp.

Chapter 2

1. There are two reports known to us of dinosaur footprints indicating maimed animals that had lost a part of one toe. Tucker, M. E., and Burchette T. P. 1977. "Triassic Dinosaur Footprints from South Wales; Their Context and Preservation." *Paleogeography, Paleoclimatology, and Paleoecology* v. 22, p. 197. Abel, O. 1935. Vorzeitliche Lebensspuren. Jena: Gustav Fisher.
2. This conclusion was reached by R. McNeill Alexander in 1985. He suggested that brontosaurs carried up to 80 percent of their weight on their hind feet. This interpretation is supported by the simple observation that the back footprints average about four times the surface area of the front footprints. Alexander, R. M. 1985. "Mechanics of Posture and the Gait of Some Large Dinosaurs." *Zoological Journal of the Linnaean Society* v. 83, pp. 1–25. Alexander, R. M. 1989. *Dynamics of Dinosaurs and Other Extinct Giants*. New York: Columbia University Press, 167 pp.
3. See discussion in Chapter 13.
4. Unfortunately, such tracks have been a significant factor in exacerbating the so-called evolution–creationism debate. Kuban, J. 1989. "Elongate Dinosaur Tracks." In D. D. Gillette and M. G. Lockey (Eds.) *Dinosaur Tracks and Traces*. New York: Cambridge University Press, pp. 427–440.
5. Some dinosaur trackers go so far as to suggest that individual tracks are relatively useless or at least highly misleading in any analysis.

Chapter 3

1. Identifying these dynamic phases of footsteps is not easy, especially in fossil tracks; however, it has been done in a few cases. Thulborn, R., and Wade, M. 1984. "Dinosaur Trackways in the Winton Formation (Mid-Cretaceous) of Queensland." *Memoir of the Queensland Museum* v. 21, p. 413–517.
2. Because tracks represent the locomotion of animals, they represent the dynamic contact between foot and substrate rather than the simple molding of the sole. Some trackers have made molds from the feet of dead animals for comparison with fossil footprints. For example, Frank Peabody illustrated front- and hind-foot "impressions taken from a pickled specimen of *Sphenodon*," the New Zealand tuatara lizard. He then compared these with 250-million-year-old Triassic tracks. Baird, D. 1980. "A Prosauropod Trackway from the Navajo Sandstone (Lower Jurassic) of Arizona." In L. L. Jacobs, (Ed.) *Aspects of Vertebrate History*. Flagstaff: Museum of Northern Arizona Press, pp. 219–30. Peabody, F. 1948. "Reptile and Amphibian Trackways from the Lower Triassic Moenkopi Formation of Arizona and Utah." *University of California Publications Bulletin of the Department of Geological Science* v. 27 (no. 8), p. 295–468.
3. An *Iguanodon*-like footprint with skin impressions is known from the Lower Cretaceous of Colorado. A *hadrosaur* footprint with skin impressions is known from the Upper Cretaceous of Alberta. Lockley, M. G. 1988. "Dinosaurs near Denver." *Geological Society of America Centennial Fieldguide*. Colorado School of Mines. Professional Contribution no. 12.
4. Frank Peabody described a number of nondinosaurian reptile footprints with clearly defined skin or scale impressions. Peabody, F. 1948. "Reptile and Amphibian Trackways."
5. Edward Hitchcock realized the importance of undertracks, but few people have followed his example by studying them seriously. Hitchcock, E. 1858. *Ichnology of New England. A Report on the Sandstone of the Connecticut Valley, Especially Its Fossil Footmarks*. Natural Sciences of America reprint. Boston: Arno Press. 220 pp.
6. Although this was probably the exception rather than the rule, the smallness of dinosaur front feet would allow them to exert considerable force over a small area. Even if they carried less weight, they could have exerted force equal to or greater than that of hind feet in some cases. Lockley, M. G., and Conrad, K. 1989. "The Paleoenvironmental Context, Preservation and Paleoecological Significance of Dinosaur Tracksites in the Western U.S.A." In D. D. Gillette and M. G. Lockley (Eds.) *Dinosaur Tracks and Traces*. New York: Cambridge University Press, pp. 122–34.
7. The bias in favor of preserving large tracks and undertracks may explain the relative rarity of small tracks throughout the entire Age of Dinosaurs. Lockley, M. G., and Conrad, K. 1989. "The Paleoenvironmental context, Preservation and Paleoecological Significance of Dinosaur Tracksites."
8. When Paul Olsen of Columbia University suggested this, he effectively asked dinosaur trackers to think very carefully before postulating too many different species based on tracks. Olsen, P. E. 1980. "Fossil Great Lakes of the Newark Supergroup in New Jersey." In W. Manspeizer (Ed.) *Field Studies of New Jersey Geology and Guide to Field Trips*. Annual Meeting, New York State Geological Association, pp. 352–98.
9. Jack Horner is largely responsible for proving that baby dinosaurs are not rare as once

thought. Horner, J. and Gorman, J. 1988. *Digging Dinosaurs*. New York: Workman.

10. Leonardi, G. 1981. "Ichnological Rarity of Young in North East Brazil Dinosaur Populations." *Annals Acad. Brasil Ciencias*, v. 53, pp. 345–6.

Chapter 4

1. These Early Jurassic tracks later became the subject of intensive study by Edward Hitchcock, the world's first serious student of vertebrate tracks. Hitchcock, E. 1858. *Ichnology of New England. A Report on the Sandstone of the Connecticut Valley, Especially Its Fossil Footmarks*. Boston: W. White, 232 pp., 60 pls.

2. Dinosaur tracks have been reported from several dozen coal mines and appear to be a common by-product of the coal industry in certain areas. Peterson, W. 1924. "Dinosaur Tracks in the Roofs of Coal Mines." *Natural History*, v. 24, p. 388–91. Parker, L. R., and Balsley, J. K. 1989. "Coal Mines as Localities for Studying Dinosaur Trace Fossils." In Gillette, D. D., and Lockley, M. G. (Eds.) *Dinosaur Tracks and Traces*, New York: Cambridge University Press, pp. 353–9.

3. This site, which is accessible to visitors, is discussed further in the section on dinosaur tracksites. Ostrom, J. 1968. "The Rocky Hill Dinosaurs." In *Guidebook for Fieldtrips in Connecticut*, no. 2. Connecticut Geological and Natural History Survey, pp. 1–12.

4. Despite being the largest continuously mapped dinosaur tracksite in the United States and the first to reveal brontosaur tracks, it was not seriously studied between 1935 and the 1980s. The main reason was probably its remote location. Lockley, M. G., Houck, K., and Prince, N. K. 1986. "North America's Largest Dinosaur Tracksite: Implications for Morrison Formation Paleoecology." *Geological Society of America Bulletin*, v. 57, p. 1163–76.

5. The site was rediscovered by Scott Madsen, who reported on it briefly in 1986. To this date it has still not been systematically studied. Madsen, S. 1986. "The Rediscovery of Dinosaur Tracks near Cameron, Arizona." In Gillette, D. D. (Ed.) *Abstracts with Program. First International Symposium on Dinosaur Tracks and Traces*, Albuquerque; New Mexico Museum of Natural History, p. 20.

6. Dingman, R. J., and Galli, C. O. 1965. "Geology and Groundwater Resources of the Pica Area. Tarapaca Province, Chile." *Geological Survey Bulletin*, no. 1189, pp. 1–113.

7. When the site was studied by a less outspoken scientist, the "man tracks" were shown to be dinosaurian in origin. Kuban, G. J. 1989.

"Elongate Dinosaur Tracks," In Gillette, D. D., and Lockley, M. G. (Eds.) *Dinosaur Tracks and Traces*. New York: Cambridge University Press, pp. 57–72.

8. Pitman, J. G., and Gillette, D. D. 1989. "The Briar Site: A New Sauropod Dinosaur Tracksite in Lower Cretaceous beds of Arkansas, USA." In Gillette, D. D., and Lockley, M. G. (Eds.) *Dinosaur Tracks and Traces*. New York: Cambridge University Press, pp. 311–32.

9. The author uses a single site documentation sheet to summarize these vital statistics at each known site being studied.

10. The map of a site at Alameda Parkway near Denver makes immediately clear the exact location of trackways made by four-footed herbivores and two-footed carnivores. Lockley, M. G. 1987. "Dinosaur Footprints from the Dakota Group of Eastern Colorado." *Mountain Geologist*, v. 24, p. 107–22.

11. The excavation was also time consuming and costly in terms of labor intensiveness. The story, which makes fascinating reading, is discussed here in Chapter 13, as well as in R. T. Bird's autobiography. Bird, R. T. 1985. *Bones for Barnum Brown. Adventures of a Dinosaur Hunter*. Fort Worth: Texas Christian University Press, 225 pp.

12. Jeffrey Pittman has used this prediction method to discover several new tracksites in Cretaceous track-bearing strata in Texas. The author has had similar success in tracing sites along the outcrop of two similar, extensive track-bearing zones in Colorado and Utah. Pittman, J. G. 1989. "Stratigraphy, Lithology, Depositional Environment and Track Type of Dinosaur Track-Bearing Beds of the Gulf Coastal Plain." In Gillette, D. D., and Lockley, M. G. (Eds.) *Dinosaur Tracks and Traces*. New York: Cambridge University Press, pp. 135–53. Lockley, M. G., Conrad, K., and Jones, M. 1988. "Regional Scale Dinosaur Track-Bearing Beds: New Tools for Sedimentologists and Stratigraphers." *Geological Society of America* (Abstracts with Program), v. 20, p. 316.

Chapter 5

1. *Brasilichnium* was named by Giuseppe Leonardi in 1980 to describe possible mammal tracks from the age of dinosaurs. Leonardi, G. 1980. "Novo Ichnogenero de Tetrapode Mesozoico da Formacao Botucatu, Avaraquara, sp." *Anals Acad. Brasil. Cienc.*, v. 53, pp. 793–805.

2. The latest and most thorough compilation of dinosaur species names lists about 900 based on bones and about 570 based on tracks. Although there is some unnecessary duplication in names of skeletal remains, in the case of

tracks the number is considerably inflated by the unfortunate practices of giving several names to one type of footprint. Chure, D. J., and McIntosh, J. S. 1989. *A Bibliography of the Dinosauria Exclusive of the Aves 1677–1986.* Paleontology Series no. 1. Grand Junction: Museum of Western Colorado, 226pp.

3. The bipedal is considered the ancestral condition.

4. Recently the traditional subdivision of carnivorous dinosaurs into Carnosauria and Coelurosauria has been reevaluated to produce a new classification with three groups. In this new classification the Ceratosauria are proposed as a third group. However, this classification has yet to be widely adopted outside specialist scientific circles. Rowe, T. 1989. "The Early History of Theropods." In Padian, K. (Ed.) *The Age of Dinosaurs.* Short Course no. 2. Knoxville, Tenn.: Paleontological Society, pp. 100–12.

5. In his classic 1858 monograph Edward Hitchcock named what is still one of the best-known dinosaur tracks, *Grallator* (one who goes on stilts). He surmised: "[T]he number and position of the toes ally them to certain kinds of birds." Hitchcock, E. 1858. *Ichnology of New England: A Report on the Sandstone of the Connecticut Valley and Its Fossil Footmarks.* Boston, W. White 220 pp.

6. Lull, R. S. 1904. "Fossil Footprints of the Jura-Trias of North America." *Memoirs of the Boston Museum of Natural History,* v. 5, pp. 461–557. See also Baird, D. 1980. "A Prosauropod Trackway from the Navajo Sandstone (Lower Jurassic) of Arizona." In Jacobs, L. L. (Ed.) *Aspects of Vertebrate History.* Flagstaff: Museum of Northern Arizona Press, pp. 219–30.

7. Hitchcock, E. 1845. "An Attempt to Name, Classify and Describe the Animals That Made the Fossil Footmarks of New England." *Sixth Annual Meeting of the Association of American Geologists and Naturalists,* pp. 23–5.

8. Sternberg, C. 1926. "Dinosaur Tracks from the Edmonton Formation of Alberta." *Geological Survey of Canada Bulletin,* v. 44, pp. 85–7.

9. Haubold, H. 1971. "Ichnia Amphibiorum et Reptiliorum Fossilium." *Handbuch der Paleoherpetologie,* part 18. Gustav Fischer, p. 121. Peterson, W. 1924. "Dinosaur Tracks on the Roof of Coal Mines." *Natural History,* v. 24, p. 388. The original track and trackway was reported, but not named, by Peterson.

10. *Otozoum* was first described by Hitchcock, who named the tracks *Otozoum moodi* in honor of "Pliny Moody Esq., of South Hadley" who was the first man "who discovered and preserved as objects of interest, fossil foot marks,

near the beginning of the present [19th] century." Hitchcock, "Ichnology of New England," p. 125. Moody's son, Plinius Moody, later pointed out the impressions of the forefeet, which appear to be preserved only occasionally. From this Hitchcock concluded that the animal was mainly bipedal. In 1904 Lull referred the *Otozoum* trackmaker to prosauropods, a position followed by Haubold as recently as 1971. Don Baird, however, considers that the trackmaker may have been a crocodilian. This position has since been adopted by Haubold and by Padian and Olsen. It appears the jury is still out with regard to the affinity of the *Otozoum* trackmaker. Hitchcock, E. 1847. "Description of Two New Species of Fossil Footmarks Found in Massachusetts and Connecticut or of the Animals That Made Them." *American Journal of Science,* v. 2, pp. 46–7. Padian, K. 1986. *The beginning of the Age of Dinosaurs.* New York: Cambridge University Press.

11. Farlow, J. O., Pittman, J. G., and Hawthorne, M. 1989. *Brontopodus birdi* Lower Cretaceous Sauropod Footprints from the U.S. Gulf Coastal Plain." In Gillette, D. D., and Lockley, M. G. (Eds.) *Dinosaur Tracks and Traces.* New York: Cambridge University Press, pp. 371–94.

12. Hitchcock, E. 1848. "An Attempt to Discriminate and Describe the Animals That Made the Fossil Footmarks of the United States, and Especially of New England." *Transactions of the American Academy of Arts and Science (N.S.),* v. 3, pp. 129–256.

13. Sternberg. "Dinosaur Tracks from the Peace River, British Columbia."

14. Leonardi, G. 1984. "Le impronte fossili de dinosauri." In Bonaparte, J. F., Colbert E. H., Currie, P. J., et al. *Sulle orme die dinosauri* Venice: Erizzo Editrice, pp. 165–86. See also Lockley, M. G. 1987. "Dinosaur Footprints from the Dakota Group of Eastern Colorado." *Mountain Geologist,* v. 24, p. 107.

15. Casamiquela, R., and Fasola, A. 1968. "Sobre pisadas de dinosaurios del Cretacio inferior del Colchagua (Chile)." *Chile Univ. Cienc. Fis. Mat. Dept. Geol. Publ.,* v. 30. The so-called *Iguanodon* tracks have remained a fascination for scientists and lay persons alike. Sir Arthur Conan-Doyle is said to have received inspiration for *The Lost World* from *Iguanodon* and *Iguanodon* track discoveries in England. Batory, D. R., and Sarjeant, W. A. S. 1989. "Sussex Iguanodon Footprints and the Writing of *The Lost World.*" In Gillette, D. D., and Lockley, M. G. (Eds.) *Dinosaur Tracks and Traces.* New York: Cambridge University Press, pp. 13–18.

16. The name *Hadrosaurichnus* was coined by Alonso in 1980 to describe a presumed hadrosaur track from Argentina. Alonso, R. 1980. "Ichnites de dinosaurios (Ornthopoda, Hadrosauroidae) en el Cretacio superior del norte Argentina." *Acta Geologica Lilloana*, v. 15, no. 2, pp. 55–63.

17. The term *quadrupedal ornithischian* distinguishes plated, armoured, and horned dinosaurs (stegosaurs, ankylosaurs, and ceratopsians) from the ornithopods.

18. Sternberg, "Dinosaur Tracks from the Peace River, British Columbia." Unfortunately the trackway was flooded by the damming of Canada's Peace River in the 1970s.

19. The nodosaurid ankylosaur interpretation, proposed by Ensom, probably represents the best interpretation of the tracks. Ensom, P. 1987. "Dinosaur Tracks in Dorset." *Geology Today*, pp. 182–3.

20. This is one of the few segments of trackway that has been confidently assigned to a ceratopsian trackmaker. Lockley, M. G. 1986. "Dinosaur Tracksites." *University of Colorado at Denver, Geology Department Magazine*, Special Issue no. 1, 56pp.

Chapter 6

1. Coombs based his classification of dinosaur running ability on consideration of dinosaur anatomy. Coombs, W. P. 1978. "Theoretical Aspects of Cursorial Adaptations in Dinosaurs." *Quarterly Review of Biology*, v. 53, p. 393.

2. This may also be the only reliable, documented example of a large running dinosaur reported to date. Farlow, J. O. 1981. "Estimates of Dinosaur Speed from a New Trackway Site in Texas." *Nature*, v. 294, p. 747.

3. Alexander's formula is as follows:
    Speed $= 0.25g^{0.5} \times SL^{1.67} \times h^{-1.17}$
    where $g =$ acceleration due to gravity, $SL =$ stride length, and h (hip height) $= 4 \times$ foot length. Alexander, R. McN. "Estimates of Speeds of Dinosaurs." *Nature*, v. 261, pp. 129–30. Thulborn and others replied with comments and rebuttals (see Chapter 13).

4. Thulborn, R. A. 1982. "Speeds and Gaits of Dinosaurs." *Paleogeography, Paleoclimatology, Paleoecology*, v. 38, p. 227. Thulborn R. A. 1984. "The Preferred Gaits of Dinosaurs." *Alcheringa*, v. 8, p. 243. Thulborn, R. A. 1989. "The Gaits of Dinosaurs." In Gillette, D. D., and Lockley, M. G. (Eds.) *Dinosaur Tracks and Traces*. New York: Cambridge University Press, pp. 39–50.

5. Bakker's contribution to the dinosaur speed debate is confusing and controversial, to say the least. The most glaring omission is his failure to cite the substantial, up-to-date work published by Thulborn. Bakker, R. T. 1987. "The Return of the Dancing Dinosaur." In Czerkas, S. and Olsen E. C. (Eds.) *Dinosaurs, Past and Present*. Los Angeles County Museum, 2 vols.

6. Thulborn, "The Gaits of Dinosaurs."

7. Paul, G. 1988. *Predatory Dinosaurs of the World*. New York: Simon and Schuster, 464 pp.

8. Alexander, R. McN. 1991. "How Dinosaurs Ran." *Scientific American*, v. 261 pp. 130–6.

9. Currie, P. J. 1989. "Dinosaur Footprints of Western Canada." In Gillette, D. D., and Lockley, M. G. (Eds.) *Dinosaur Tracks and Traces*. New York: Cambridge University Press, pp. 293–300.

10. Lockley, M. G. 1990. "Tracking the Rise of Dinosaurs in Eastern Utah." *Canyon Legacy*, v. 2, pp. 2–8.

11. Bird's interpretation of a small dinosaur sitting out the storm has not been confirmed by a detailed study of the track-bearing slab. Bird, R. T. 1985. *Bones for Barnum Brown: Adventures of a Dinosaur Hunter*. Fort Worth: Texas Christian University Press, 225pp.

12. There are other examples of dinosaurs taking alternating long and short steps, but none is as pronounced as the Moroccan example. Ishigaki, S. 1986. "Dinosaur Footprints of the Atlas Mountains." *Nature Study* (Japan) v. 32(1), pp. 6–9.

13. Abel, O. 1935. Vorzeitliche Lebensspuren. Jena: Gustav Fisher.

14. The validity of Bird's interpretation has yet to be adequately tested. Bird, R. T. 1944. "Did Brontosaurus Ever Walk on Land?" *Natural History*, v. 53, p. 61.

15. Unlike Bird, Ishigaki discovered several trackways with different step patterns. He appears to have been influenced by Bird in his interpretation of these as trackways of swimmers. Ishigaki, S. 1989. "Footprints of Swimming Sauropods from Morocco." In Gillette, D. D., and Lockley, M. G. (Eds.) *Dinosaur Tracks and Traces*. New York: Cambridge University Press, pp. 83–6.

16. Coombs made this point strongly in his paper. Essentially he debunked the notion that carnivores could not swim or chase prey into the water. Coombs, W. P., Jr. 1980. "Swimming Ability of Carnivorous Dinosaurs." *Science*, v. 207, p. 1198.

17. Lockley, M. G. 1990. "Did Brontosaurus Ever Swim out to Sea? Evidence from Brontosaur and Other Dinosaur footprints." *Ichnos*, v. 1, pp. 81–90. Lockley, M. G., and Conrad, K. 1989. "The Paleoenvironmental Context, Pres-

ervation and Paleoecological Significance of Dinosaur Tracksites in the Western USA." In Gillette, D. D., and Lockley M. G. (Eds.) *Dinosaur Tracks and Traces*. New York: Cambridge University Press, pp. 121–34.

Chapter 7
1. Bird, R. T. 1941. "A Dinosaur Walks into the Museum." *Natural History*, v. 47, pp. 75–81. Bird, R. T. 1944. "Did Brontosaurus Ever Walk on Land?" *Natural History*, v. 53, pp. 61–7. The evidence for a herd is presented here, but an analysis of "herd structure" is presented in Chapter 13.
2. Ostrom's discussion was based entirely on footprint evidence. Ostrom, J. 1972. "Were Some Dinosaurs Gregarious?" *Paleogeography, Paleoclimatology, Paleoecology*, v. 11, pp. 287–301.
3. The Texas tracks were originally interpreted as ornithopod tracks. Albritton, C. L. 1942." Dinosaur Tracks Near Comanche, Texas." *Field and Lab*, v. 10, p. 160.
4. Each animal requires its own space to avoid collision with its nearest neighbors. Gregarious birds and mammals often travel at regularly spaced intervals either on a broad front, in a file or line, or in an intermediate *V*, or spearhead, formation. Lockley, M. G. 1989. "Tracks and Traces: New Perspectives on Dinosaur Behavior, Ecology, and Biogeography." In Padian, K. (Ed.) *The Age of Dinosaurs*. Short Course no. 2. Knoxville, Tenn.: Paleontological Society, pp. 134–145. Cohen, A., Lockley, M. G., Halfpenny, J., and Michel, A. E. 1990. *Modern Vertebrate Track Formation and Preservation at Lake Manyara, Tanzania: Implications for Paleoecology and Paleobiology*. National Geographic Open File report.
5. Currie, P. J. 1983. "Hadrosaur Trackways from the Lower Cretaceous of Canada." *Acta Paleontologica Polonica*, v. 28, pp. 62–74.
6. The attack scenario is analyzed in Chapter 13.
7. This analysis, presented in Chapter 13, shows how track overlap patterns define a sequence of events in time. Lockley, M. G. 1987. "Dinosaur Trackways." In Czerkas, S., and Olsen, O. C. (Eds.) *Dinosaurs, Past and Present*. Los Angeles: Los Angeles County Museum, pp. 81–95.
8. Seen in Figure 7.3; see Lockley, M. G., Houck, K., and Prince, N. K. 1986." North America's Largest Dinosaur Tracksite: Implications for Morrison Formation Paleoecology." *Bulletin of the Geological Society of America*, v. 97, p. 1163.
9. Albritton, C.L. 1942 "Dinosaur Tracks Near Commanche, Texas." *Field and Lab*, v. 10.
10. Many studies of bone remains indicate only a small proportion of carnivores. Low predator/prey ratios provide one of the cornerstones of the warm-blooded dinosaur theory. Ostrom, J. H. 1969. "Terrestrial Vertebrates as Indicators of Mesozoic Climates." *North American Paleontological Convention Proceedings, Pt. D.*, 347–76. Bakker, R. T. 1975. "Dinosaur Renaissance." *Scientific American*, v. 232, p. 58.
11. This appealing idea has been persistently quoted and requoted despite a lack of convincing evidence in the track record. In 1989 *National Geographic* magazine published a poster of an *Ultrasaurus* herd protecting juveniles from attacking carnivores. Steven Jay Gould also talked of "a herd of migrating ornithopods, with vulnerable juveniles in the center and strong adults at the peripheries." Nowhere in the world is there any trackway evidence for either of these scenarios. Bakker, R. T. 1968. "The Superiority of the Dinosaurs." *Discovery*, v. 2, p. 11. Gould, S. J. 1989. "The Dinosaur Rip-off." *Natural History*, p. 14–18.
12. This is probably the best evidence anywhere to suggest that theropods sometimes roamed and hunted in packs. Leonardi, G. 1984. "Le Impronte Fossili de Dinosauri." In Bonaparte, J. F. et al. (Eds.) *Sulle Orme dei Dinosauri*. Venice: Erizzo Editrice, pp. 165–86.
13. Thullborn, R. A., and Wade, M. 1979. Dinosaur Stampede in the Cretaceous of Queensland. *Lethaia*, v. 12, p. 275. Thulborn, R. A., and Wade, M. 1979. "Dinosaur Trackways in the Winton Formation (Mid Cretaceous) of Queensland." *Memoirs of the Queensland Museum*, v. 21, pp. 413–517.
14. Paul, G. 1988. *Predatory Dinosaurs of the World*. New York: Simon and Schuster, 464pp.
15. Several authors, including Giuseppe Leonardi and James Farlow, have suggested that activity levels varied considerably between different dinosaur species (see Bibliography). Farlow, J. O. 1987. "Lower Cretaceous Dinosaur Tracks, Paluxy River Texas" *Geological Society of America, South Central Magazine*, 50pp.

Chapter 8
1. There are regularly held conferences on Mesozoic terrestrial ecosystems (see note 6).
2. Paleontologist traditionally regard bones as more useful than tracks because one can identify the animal at the species level and not just at the family or order level.
3. For example the 40 percent sauropod category may include 20 percent type A and 20 percent type B. This situation is similar to that recorded at the Late Jurassic Purgatoire site in Colorado. Lockley, M. G., Houck, K., and Prince, N. K. 1986. "North America's Largest

Dinosaur Tracksite: Implications for Morrison Formation Paleoecology." *Geological Society of America Bulletin*, v. 97, pp. 1163–76. Lockley, M. G., and Price, N. K. 1988. "The Purgatoire Valley Dinosaur Tracksite Region, Southeast Colorado." *Geological Society of America Fieldtrip Guidebook*. Professional Contributions. Golden: Colorado School of Mines, pp. 275–87.

4. This scheme was introduced by the author in 1989 to help dispel the prevalent notion that tracks are only useful where bones are rare or absent. Lockley, M. G. 1989. "Tracks and Traces: New Perspectives on Dinosaurian Behavior, Ecology and Biogeography." In Padian, K., and Chure, D. (Eds.) "The Age of Dinosaurs." *Paleontological Society Short Course no. 2*, pp. 134–45.

5. A long-awaited monograph on *Coelophysis* has recently been published. Colbert, E. H. 1989. "The Triasssic Dinosaur *Coelophysis*." *Museum of Northern Arizona Press Bulletin*, v. 57, 160pp. See also Long, R. A., Houk, R. 1988. *Dawn of the Dinosaurs: The Triassic in Petrified Forest*. Petrified Forest Museum Association, 96pp.

6. The author first published a dinosaur community census based on footprints in 1987, using the Popo Agie (pop-OWE-jah) as an example. Giuseppe Leonardi also used similar track census methods in his studies of South American tracks. Lockley, M. G., and Conrad, K. 1987. "Mesozoic Tetrapod Tracksites and Their Application in Paleoecological Census Studies." In Currie, P., and Koster, E. H. (Eds.) *Fourth Symposium on Mesozoic Terrestrial Ecosystems*. Drumheller, Canada, pp. 144–9.

7. Because the Early Jurassic is known as the Liassic and because the tracks from this time are so classic, I have referred to them as classic Liassic footprint assemblages. See Chapters 5 and 7 for further discussion. Lockley, M. G. 1989." Classic Liassic Footprint Assemblages from the Colorado Plateau." *Journal of Vertebrate Paleontology*, v. 9, pp. 29A–30A.

8. The rift valley was formed, in much the same way as the present East African rift, when Europe and the Old World began to split apart from the Americas. Water was in ample supply, as shown by lake deposits. The environment was quite lush, and as shown by the tracks, animals were plentiful.

9. William Stokes referred to "a desert as a place almost barren of life" but then, on the basis of track evidence, concluded that "dinosaurs were fairly abundant." Samuel Welles also noted that tracks were abundant in these desert deposits and observed: "A complete study of the footprints . . . would repay the effort involved." Stokes, W. L. 1978. "Animal Tracks

in the Navajo-Nugget sandstone." *Contributions to Geology, University of Wyoming*, v. 16, pp. 103–7. Stokes, W. L., and Bruhn, A. F. 1960. "Dinosaur Tracks from Zion National Park and Vicinity, Utah." *Proceedings of the Utah Academy of Science, Arts and Letters*, v. 37, pp. 75–6. Welles, S. P. 1970. "Dinosaur Footprints from the Kayenta Formation of Northern Arizona." *Plateau*, v. 44, pp. 27–38.

10. The author has recently discovered a large number of mammal-like trackways attributable to *Brasilichnium* (see Chapter 5).

11. The history of discovery in this area is briefly outlined in English by Shinobu Ishigaki. There is also a significant amount of French and Japanese literature. Ishigaki, S. 1989. "Footprints of Swimming Sauropods from Morocco." In Gillette, D. D., and Lockley, M. G. (Eds.) *Dinosaur Tracks and Traces*. New York: Cambridge University Press, pp. 83–6.

12. Ostrom, J., and McIntosh, J. 1966. *Marsh's Dinosaurs*. New Haven: Yale University Press.

13. Lockley and Price, "The Purgatoire Valley Dinosaur Tracksite Region."

14. Paleontologists studying the Morrison formation usually ask us to imagine a Serengeti-like African savannah when we try to visualize the ancient environment inhabited by Morrison dinosaurs.

15. It is only recently that any distinct theropod tracks have been reported. Woodhams, K. E., and Hines, J. S. 1989. "Dinosaur Footprints from the Lower Cretaceous of East Sussex, England." In Gillette, D. D., and Lockley, M. G. (Eds.) *Dinosaur Tracks and Traces*. New York: Cambridge University Press, pp. 301–7.

16. Godoy, L. C., and Leonardi, G. 1985. "Direcoes e comportamento dos dinosauros de localidad de Piau, Sousa, Paraiba (Brasil), formacao Sousa (Cretaceo Inferior)." Department Nacional de Producao Mineral, Coletanea de Trabalhos Paleontologicos, *Serie Geologia*, v. 27, pp. 65–73. Leonardi, G. 1984." Le impronte fossili de dinosauri sulle orme dei dinosauri." In Bonaparte, J., et al. (Eds.) *Sulle orme dei dinosauri*. Venice: Erizzo Editrice, pp. 165–86.

17. The trackmakers were originally interpreted as hadrosaur, or duck-billed, dinosaurs. This trackmaker identification has been disputed because the deposits are older than any known to contain hadrosaur remains. Instead, the tracks were probably made by *Iguanodon*-like species, ancestors of the true duck-bills. Currie, P. J. 1983. "Hadrosaur Trackways from the Lower Cretaceous of Canada." *Acta Paleontologica Polonica*, v. 28, pp. 63–73.

18. There are a handful of early Cretaceous re-

ports of true bird tracks associated with deposits of this type. Currie, P. J. 1981. "Bird Footprints from the Gething Formation (Lower Cretaceous) of Northeastern British Columbia, Canada." *Journal of Vertebrate Paleontology*, v. 1, pp. 257–64.

19. There are a number of studies dealing with dinosaur tracks in the Dakota Group. Lockley, M. G. 1987. "Dinosaur Footprints from the Dakota Group of Eastern Colorado." *Mountain Geologist*, v. 24, pp. 107–22. Lockley, M. G., Matsukawa, M., and Obata, I. 1989. "Dinosaur Tracks and Radial Cracks: Unusual Footprint Features." *Bulletin of the Natural Science Museum Tokyo*, ser. C, pp. 151–60.

20. These environments were similar to the limy coastal plains described for the Middle Jurassic of Morroco. Pittman, J. 1989. "Stratigraphy, Lithology, Depositional Environment and Track Type of Dinosaur Trackbearing Beds of the Gulf Coastal Plain." In Gillette, D. D., and Lockley, M. G. (Eds.) *Dinosaur Tracks and Traces*. New York: Cambridge University Press, pp. 135–54.

21. As discussed in Chapter 10, this makes the Jindong beds among the most extensively tracked of all deposits from the Age of Dinosaurs. Lim, S. K., Yang, S. Y., and Lockley, M. G. 1989. "Dinosaur Tracks from the Jindong Formation of Korea." In Gillette, D. D., and Lockley, M. G. (Eds.) *Dinosaur Tracks and Traces*. New York: Cambridge University Press, pp. 333–6.

22. According to Greg Paul, it is possible that all the tracks represent theropods (see Chapter 7). If this is the case, the Winton site cannot be used to provide a reliable paleoecological census.

23. Even after the rush of dinosaur exploration to these formations in the early decades of the twentieth century, very few tracks were reported. Lockley, M. G., Young, B. H., and Carpenter, K. 1983. "Hadrosaur Locomotion and Herding Behavior: Evidence from Footprints in the Mesa Verde Formation, Grand Mesa Coalfield, Colorado." *Mountain Geologist*, v. 20, pp. 5–13.

24. These tracks are often preserved in coal mines. (See Chapter 4 and note 23).

25. Tracks from the Laramie Formation were reported by the author, first in 1986 and again in 1988. Lockley, M. G. 1988. "Dinosaurs Near Denver." *Geological Society of American Field Guide*. Colorado School of Mines Professional Contribution no. 12. Recent field work has established the presence of many track-bearing layers.

26. This track assemblage resembles that seen in the Late Jurassic. See notes 3 and 16.

27. John Ostrom independently reached the conclusion that such humid, well-vegetated coastal plain environments were the preferred habitat of duck-billed dinosaurs. Ostrom, J. H. 1964. "A Reconsideration of the Paleoecology of Hadrosaurian Dinosaurs." *American Journal of Science*, v. 262, pp. 975–97.

28. We have pointed out this distinction in a recent paper. We believe that the understanding of the relationship between track type and ancient environment is critical for any dinosaur census studies. Lockley, M. G., and Conrad, K. 1989. "The Paleoenvironmental Context, Preservation and Paleoecological Significance of Dinosaur Tracksites in the Western USA." In Gillette, D. D., and Lockley, M. G. (Eds.) *Dinosaur Tracks and Traces*. New York: Cambridge University Press, pp. 121–34.

29. This quote is attributed to the ecologist Edward S. Deevey.

Chapter 9

1. The carnivorous theropods belong to the saurischian group of dinosaurs, whereas the herbivorous ornithopods belong to the ornithischian group.

2. As discussed in Chapter 5, *Tetrapodosaurus* tracks were first attributed to a horned dinosaur, then later to an armored dinosaur. Similarly newly found "sauropod" tracks may be attributable to an armored dinosaur.

3. Their papers on this subject include one whose title loosely translates as "The Bearing of Triassic Tracks on the Problem of the Origin of Dinosaurs." Demathieu, G., and Haubold, H. 1978. "Du problem de l'origine des dinosauriens d'après les données de l'ichnologie du Trias." *Geobios*, v. 11 (3), pp. 409–12. Demathieu, G. 1989. "Appearance of the First Dinosaur Tracks in the French Middle Triassic and their Probable Significance." In Gillette, D. D., and Lockley, M. G. (Eds.) *Dinosaur Tracks and Traces*. New York: Cambridge University Press, pp. 201–7.

4. In North America there is a very good record of Triassic and Jurassic theropod tracks, so we can be fairly sure that the maximum-size trend is a fair reflection of this aspect of carnivore evolution. In theory we could do the same for other trackmaking groups. For example, the maximum size of brontosaur tracks also tends to increase during the Jurassic. In practice, the track record may not be complete enough to do this for every group.

5. *Atreipus* is a track type not yet introduced. As indicated by Figure 9.3 it is the oldest (stratigraphically lowest) track type attributed to an ornithopod. Olsen, P. E., and Baird, D. 1986. "The Ichnogenus *Atreipus* and Its significance for Triassic Biostratigraphy," In Padian, K. (Ed.) *The Beginning of the Age of Dinosaurs*. New York: Cambridge University Press, pp. 61–87.

6. This quote is attributed to Robert Bakker. Bakker, R. T. 1978. "Dinosaur Feeding Behavior and the Origin of Flowering Plants." *Nature*, v. 274, p. 661–3.

7. As discussed in Chapter 2, sauropod trackways show that they walked very erectly. This evidence has helped revolutionize reconstructions of sauropod skeletons.

8. Robert Bakker's dancing dinosaurs, including the bipedal stegosaur, were discussed in Chapter 6. His ideas have resulted in bipedal restorations, for example, by Dinamation Inc. Bakker, R. T. 1987. "The Return of the Dancing Dinosaur." In Czerkas, S., and Olsen, E. C. (Eds.) *Dinosaurs Past and Present*. Los Angeles: Los Angeles County Natural History Museum, pp. 37–79.

9. Chure, D. 1989. "Quo Vadis *Tyrannosaurus?*: The Future of Dinosaur Studies." In Padian, K., and Chure, D. (Eds.) *The Age of Dinosaurs*. Short Courses in Paleontology, no. 2. Knoxville, Tenn.: Paleontological Society, pp. 175–83.

10. Flowering plants had not yet evolved.

11. Certain Cretaceous sauropod communities have been identified as "inland" communities, associated with drier, upland conditions. Lucas, S. G. 1981. "Dinosaur Communities of the San Juan Basin; a Case for Lateral Variations in the Composition of Late Cretaceous Dinosaur Communities." In Lucas, S. G., Kues, B. S., and Rigby, J. K. (Eds.) *Advances in San Juan basin Paleontology*. Albuquerque: University of New Mexico Press, pp. 337–93.

12. Only stomach contents can prove what these dinosaurs ate, and such remains are extremely rare. The abundance of ornithopod tracks in environments dominated by flowering plants, however, makes it probable that dinosaurs ate this type of plant.

13. A good review of this topic is provided by Gauthier, J. A., and Padian, K. 1989. "The Origin of Birds and Evolution of Flight." In Padian, K., and Chure, D. (Eds.) *The Age of Dinosaurs*. Short Course no. 2. Knoxville, Tenn.: Paleontological Society, pp. 121–33.

14. Paul Sereno reported the remains of an Early Cretaceous bird from China that is more advanced than *Archaeopteryx* and Sankar Chatterjee of Texas Tech. reported the remains of *Protoavis*. Both paleontologists are working on detailed descriptions of the material.

15. Cretaceous bird tracks were first reported from the Dakota Group of Colorado in 1931. Another well-documented find was made in Canada in 1980. Since then tracks have been found in China, Japan, and South Korea. (See Chapter 8, notes 18 and 19.)

16. The North American birdlike tracks are found in the Navajo Formation of Eastern Utah. The North African tracks, from Morocco, were described by Ishigaki. Ishigaki, S. 1985. "Dinosaur Footprints of the Atlas Mountains." *Nature Study*, v. 31, pp. 5–7.

Chapter 10

1. Lockley, M. G. "The Paleobiological and Paleoenvironmental Importance of Dinosaur Tracks." *Palaois*, v. 1, pp. 37–47.

2. Geologists determine the patterns of the motion of tectonic plates by using magnetic studies to estimate the positions of the continents relative to the North and South poles at different times in earth history.

3. In Greek mythology Tethys was the wife of the god Oceanus. *Tethys* is also the Greek word for "oyster" and the scientific name for a family of marine snails.

4. From Late Carboniferous and Permian times through the Triassic to Early Jurassic many communities have broad or worldwide distributions.

5. Mammal-like trackmakers were not confined to desert environments for the entire 100-million-year duration from Late Carboniferous to Early Jurassic. However, they are characteristic of many desert dune deposits at this time. They also appear to survive in these environments after they have disappeared or declined in other environments.

6. Iguandon-like dinosaur tracks were reported from the island of Spitzbergen in the early 1960s. This area, currently well north of the Arctic Circle, was also at high latitude during the Cretaceous. Lapparent, A. F. de. 1962. "Footprints of Dinosaurs in the Lower Cretaceous of Vestspitzbergen-Svalbard." *Arbok/Norsk Polaristitut* (1960), pp. 14–21.

7. The water table is simply the level at which the ground becomes saturated with water.

8. The term *bioturbation* means disturbance of sediments by organisms. In many cases it simply means burrowing, but it also refers to trampling by vertebrates.

9. The term *dinotorbation* was introduced in Dodson, P., Behrensmeyer, A. K., Bakker, R. T.,

and McIntosh, J. S. 1980. In "Taphonomy and Paleoecology of the Dinosaur Beds of the Jurassic Morrison Formation." *Paleobiology*, v. 6, pp. 208–32.

10. The Triassic–Jurassic sedimentary cycles of this region have been intensively studied, and it has been shown that tracks occur regularly and at predictable levels in the strata. Olsen, P. E. 1886. "A 40-Million Year Lake Record of Early Mesozoic Orbital Climatic Forcing." *Science*, v. 234, pp. 842–8.

11. This study was a landmark in recognition of the impact large animals have on modern and ancient substrates. Laporte, L., and Behrensmeyer, A. K. 1980. "Tracks and Substrate Reworking by Terrestrial Vertebrates in Quaternary Sediments of Kenya." *Journal of Sedimentary Petrology*, v. 50, pp. 1337–46.

12. Pittman, J. G. 1989. "Stratigraphy, Lithology, Depositional Environment and Track Type of Dinosaur Track-Bearing Beds of the Gulf Coastal Plain." In Gillette, D. D., and Lockley, M. G. (Eds.) *Dinosaur Tracks and Traces*. New York: Cambridge University Press, pp. 135–51.

13. *Paleogeography* means "ancient geography"; it is a well-established field in the earth sciences.

14. Lockley, M. G., and Rice, A. 1980. "Did *Brontosaurus* Ever Swim Out to Sea?. Evidence from Brontosaur Trackways and Other Dinosaur Footprints." *Ichnos*, v. 1, pp. 81–90.

15. Gierlinski, G., and Potemska, A. 1987. "Lower Jurassic Dinosaur Footprints from Gliniany Las, Northern Slope of the Holy Cross Mountains, Poland." *Neues Jb. Geol. Palaeont. Abh.*, v. 175, pp. 107–20.

16. McAllister reported swimming ornithischian tracks from Cretaceous deposits in Kansas. Others believe these to be crocodile tracks. McAllister, J. A. 1989. "Dakota Formation Tracks from Kansas: Implications for the Recognition of Tetrapod Subaqueous Traces." In Gillette, D. D., and Lockley, M. G. (Eds.) *Dinosaur Tracks and Traces*. New York: Cambridge University Press, pp. 343–8.

17. Seilacher, A. 1986. "Footprint Is an Experiment in Soil Mechanics." In Gillette, D. D. (Ed.) *Abstracts with Program. First International Symposium on Dinosaur Tracks and Traces*. Albuquerque: New Mexico Museum of Natural History. See also Chapter 11.

18. Edwin Colbert was one of the first researchers to estimate dinosaur weight. Subsequently McNeill Alexander and others have engaged in this line of research. Colbert, E. H. 1962. "The Weights of Dinosaurs." *American Museum Noviates*, no. 2076, pp. 1–16.

19. A levee is the raised bank of a large, meandering river. Lockley, M., Matsukawa, M., and Obata, I. 1989. "Dinosaur Tracks and Radial Cracks: Unusual Footprint Features." *Bulletin of the National Science Museum, Tokyo*, series C, v. 15, pp. 151–60.

20. Brand, L. 1979. "Field and Laboratory Studies on the Coconino Sandstone (Permian) Vertebrate Footprints and Their Paleoecological Implications." *Paleogeography Paleoclimatology and Paleoecology*, v. 28, pp. 25–38.

21. Cadeé, G. C. 1990. "Feeding Traces and Bioturbation by Birds on a Tidal Flat, Dutch Wadden Sea. *Ichnos*, v. 1. pp. 23–30.

22. Boyd, D. W., and Loope, D. B. 1984. "Problematic Vertebrate Origin for Certain Sole Marks in Triassic Red Beds of Wyoming." *Journal of Paleontology*, v. 58, pp. 467–76.

23. Unwin, D. 1989. "A Predictive Method for the Identification of Vertebrate Ichnites and Its Application to Pterosaur Tracks." In Gillette, D. D., and Lockley, M. G. (Eds.) *Dinosaur Tracks and Traces*. New York: Cambridge University Press, pp. 259–74.

24. Currie, P. J. 1983. "Hadrosaur Trackways from the Lower Cretaceous of Canada." *Acta Paleontologica Polonica*, v. 28, pp. 63–73.

25. Baird, D. 1980. "A Prosauropod Trackway from the Navajo Sandstone (Lower Jurassic) of Arizona." In Jacobs, L. L. (Ed.) *Aspects of Vertebrate History*. Flagstaff: Museum of Northern Arizona Press, pp. 219–30.

26. Leonardi, G., and Godoy, L. C. 1980. "Novas pistas de tetrapodes de formacao Botucatau do estado de San Paulo." In *Anais do 31 Congr. de. Brasilero de geol. Santa Catarina*, v. 5, pp. 3080–9. McKee, E. D. 1944. "Tracks That Go Uphill." *Plateau*, v. 16, pp. 61–72.

Chapter 11

1. Ensom, P. C. 1988. "Excavations at Sunnydown Farm. Langton Matravers, Dorset: Amphibians Discovered in the Purbeck Limestone Formation." *Proceedings of the Dorset Natural History Society*, v. 109, pp. 148–50.

2. Ormund, C. 1964. *Complete Book of Outdoor Lore*. New York: Harper & Row, 498pp.

3. These weights in metric tons (= 1000 kilograms) represent average maximum weight estimates calculated by several different researchers. Alexander, R. McN. 1989. *The Dynamics of Dinosaurs and Other Extinct Giants*. New York: Columbia University Press, 186pp. Lockley, M. G., and Rice, A. 1990. "Did Brontosaurus Ever Swim Out to Sea?" *Ichnos*, v. 1, p. 81–90.

4. To date these are the only reports of dinosaurs killing invertebrates. However, this type of inadvertent dinosaur destruction must have

been quite common. Tylor. A. 1862. "On the Footprints of an *Iguanodon* Lately Found at Hastings." *Quarterly Journal of the Geological Society of London*, v. 18, pp. 247–53. Lockley, M. G., Houck, K., and Prince, N. K. 1986. "North America's Largest Dinosaur Tracksite: Implications for Morrison Formation Paleoecology." *Geological Society of America Bulletin*, v. 97, pp. 1163–76.

5. Baker, F. C. 1901. "Some Interesting Molluscan Monstrosities." *Transactions of the Academy of Sciences, St. Louis*, v. 11, pp. 143–6.

6. Similar methods are used to measure the degree of invertebrate burrowing, or bioturbation. Dodson, P., Behrensmeyer, A. K., Bakker, R. T., and McIntosh, J. 1980. "Taphonomy and Paleoecology of the Dinosaur Beds of the Jurassic Morrison Formation." *Paleobiology*, v. 6, pp. 208–32. Drosser, M., and Bottjer, D. 1986. "A Semi-Quantitative Field Classification of Ichnofabric." *Journal of Sedimentary Petrology*, v. 56, pp. 558–9.

7. Lockley, M. G., and Conrad, K. 1989. "The Paleoenvironmental Context and Preservation of Dinosaur Tracksites in the Western USA." In Gillette, D. D., and Lockley, M. G. (Eds.) *Dinosaur Tracks and Traces*. New York: Cambridge University Press, pp. 121–34.

8. Lockley, M. G. 1987. "Dinosaur Trackways." In Czerkas, S., and Olsen, E. C. (Eds.) *Dinosaurs, Past and Present. Los Angeles County Museum Symposium*. Los Angeles: Los Angeles County Museum, pp. 80–95.

9. Russell claims that large areas of homogenous Mesozoic strata, for example in the Morrison Formation, have been plowed by dinosaurs. Russell, D. A. 1989. *Dinosaurs of North America: An Odyssey in Time*. North Word Press, 239 pp.

10. Lockley, M. G., and Prince, N. K. 1988. "Dinoturbation: A Late Mesozoic Peak in the History of Trampling by Terrestrial Vertebrates." *Geological Society of America* (Abstracts with Program) v. 20, p. 377.

Chapter 12

1. It was not until 1988–9 that such extensive tracksites were first reported in the scientific literature. Lockley, M. G., and Pittman, J. G. 1989. "The Megatracksite Phenomenon: Implications for Paleoecology, Evolution and Stratigraphy." *Journal of Vertebrate Paleontology*, v. 9, p. 30A.

2. The Gulf of Mexico coastal plain provides a good modern example.

3. As indicated in Appendix A, Moab is one of the world's most interesting dinosaur footprint areas. Lockley, M. G. 1991. "The Moab

Megatracksite: A Preliminary Description and Discussion of Millions of Middle Jurassic Tracks in Eastern Utah." In Avereh, W. R. (Ed.), *Dinosaur Quarries and Tracksites*. Grand Junction Geological Society, pp. 59–65.

4. This figure represents the area where footprint sites are known. It is probable that further exploration will show that this megatracksite area is significantly larger than 300 square kilometers. Lockley and Pittman, "The Megatracksite Phenomenon." Lockley, M. G. 1989. "Tracks and Traces: New Perspectives on Dinosaurian Behavior, Ecology and Biogeography." In Padian, K., and Chure, D. J. (Eds.) *The Age of Dinosaurs*. Short Course no. 2. Knoxville, Tenn.: Paleontological Society, pp. 134–45.

5. Refractory clays are excavated for use in the brickmaking and ceramics industries. This facilitates access to good exposures of track-bearing layers.

6. The term *Dinosaur Freeway* first appeared in the scientific literature and was quickly adopted by the popular press. In Europe the term *Dinosaur Motorway* has even been used. Lockley and Pittman, "The Megatracksite Phenomenon." Jones, M. 1988. "A Dinosaur Freeway in the Cretaceous of Colorado: Implications for Stratigraphic Correlation." *Geological Society of America* (Abstracts with Program) v. 20, p. 377. Lockley, M. G. 1989. "Summary and Prospectus." In Gillette, D. D., and Lockley, M. G. (Eds.) *Dinosaur Tracks and Traces*. New York: Cambridge University Press, pp. 441–7. Lockley, M. G., Matsukawa, M., and Obata, I. 1989. "Dinosaur Tracks and Radial Cracks: Unusual Footprint Features." *Bulletin of the National Science Museum, Tokyo*, series C, v. 15, pp. 151–60.

7. Yale University's John Ostrom devoted a whole paper to demonstrating that duck-billed ornithopod dinosaurs preferred these habitats. Ostrom, J. H. 1964. "A Reconsideration of the Paleoecology of Hadrosaurian Dinosaurs." *American Journal of Science*, v. 262, pp. 975–97.

8. Pittman's study is by far the most comprehensive and innovative study of Texas tracksites since the work of Roland T. Bird over fifty years ago. Pittman, J. G. 1989. "Stratigraphy, Lithology, Depositional Environment, and Track Type of the Dinosaur Track-bearing Beds of the Gulf Coastal Plain." In Gillette, D. D., and Lockley, M. G. (Eds.) *Dinosaur Tracks and Traces*. New York: Cambridge University Press, pp. 135–54.

9. Cretaceous seas flooded into the Gulf of Mexico and northward into Colorado during a

series of pulses (transgressions and regressions). The Glen Rose represents a pulse during Albian times and the Dakota represents a subsequent Late Albian to Early Cenomanian pulse.

10. These ideas of Horner and Bakker have been around for some time. Changing sea level is thought to help explain patterns of extinction in many groups of organisms, including dinosaurs. Ginsburg, L. 1964. "Les regressions marines et le probleme du renouvellement des faunes au cours des temps géologiques." *Bulletin Société Géologique de France*, ser. 7, v. 6, pp. 13–22. Ginsburg, L. 1986. "Regression marines et extinctions des dinosaures." *Les dinosaures de la Chine à la France. Colloque International de Paleontologie*. Museum d'Histoire Naturelle de Toulouse, pp. 142–149. Horner, J. R., and Gorman, J. 1988. *Digging for Dinosaurs*. New York: Workman, 210 pp.

Chapter 13

1. The turning Jurassic brontosaur trackway and the turning Cretaceous carnivore trackway are discussed in Chapter 6.
2. Migrating African gazelles and other large mammals often show clockwise, or right-handed, patterns of migration. Grizimek, 1960. *Serengeti Shall Not Die*. London: Collins. Armen, J-C. 1974. *Gazelle Boy. L'enfant sauvage du Grand Desert*. London: Bodley Head 127 pp.
3. Charig, A. 1962. "The slow march of the Purbeck Iguanodon." *New Scientist*, no. 271, p. 186. Charig, A., and Newman, B. H. 1962. "Footprints in the Purbeck." *New Scientist*, no. 285, pp. 234–5.
4. Bernier P., et al. 1984. *"Découverte de pistes de dinosaures sauteurs dans le calcaire lithographiques de Cerin (Kimeridgian Supérieur, Ain, France): Implications paleoécologiques."* Geobios, Mémoires Specials, v. 89, pp. 177–85.
5. Thulborn, R. A. 1989. "The Gaits of Dinosaurs." In Gillette, D. D., and Lockley, M. G. (Eds.) *Dinosaur Tracks and Traces*. New York: Cambridge University Press, pp. 39–50.
6. Occam's Razor is a philosophical principle that requires common sense and simple explanations to be considered before complex and improbable hypotheses.
7. Although the author has disputed this interpretation, Bird, Farlow and others have suggested it as a valid scenario. Lockley, M. G., 1987. "Dinosaur Trackways." In Czerkas, S. and Olsen, E. C. (Eds.). *Dinosaurs, Past and Present*. Los Angeles: Los Angeles County Museum. 161 p. Bird, R. T., 1985. *Bones for Barnum Brown*. Fort Worth: Texas Christian University Press, 212 pp. For cover illustrations of the attack scene, see Farlow, J. O. 1987. "Lower Cretaceous Dinosaur Tracks, Paluxy River, Texas." *Field Guide for Geological Society of America, South Central Meeting*, 50 pp. Hastings, R. J., et al. 1990. *Dinosaur Tracks in the Cretacous Glen Rose Formation of Central Texas*. Geological Society America Field Trip no. 9.

8. Although the set of fossil footprints was made up in the twentieth century by the human animals who devised the exercise, the fact that it bears no resemblance to the fossil footprint evidence from Dinosaur State Park in Texas is not made clear. Hurd, D., et al 1989. *General Science: A Voyage of Discovery*. Englewood Cliffs, N.J.: Prentice-Hall, 577 pp.

9. The Denver Museum of Natural History houses a similar display; in this case a 70-million-year-old duck-billed dinosaur is mounted with 100-million-year-old iguanotodid tracks. A scientist there told me that, while one should not mix dinosaurs of different ages, it was okay to mix tracks!

10. Cohen, A., Lockley, M. G., Halfpenny, J., and Michel, E. 1989. "Modern Vertebrate Track Formation and Preservation at Lake Manyara, Tanzania: Implications for Paleoecology and Palebiology." National Geographic Open File Report.

11. There are examples of short traces that were probably made by dinosaur tails; however, these marks are few and far between and were not made by prolonged dragging, but by temporary touchdown of the tail. Gillette, D. D., and Thomas, D. A. 1985. "Dinosaur Tracks in the Dakota Formation (Aptian-Albian) at Clayton Lake State Park, Union County, New Mexico." In Lucas, S. G., and Zidek, J. (Eds.) *Santa Rosa Tucumcari Region, New Mexico, Geological Society Guidebook, 36th Field Conference*. Albuquerque: University of New Mexico Press, pp. 283–8.

12. Dutuit, J. M., and Quazzou, A. 1980. "Découverte d'une piste de dinosaure sauropode sur le site d'empreintes de Demnat." *Mémoire Société Géologique de France* (N.S.), v. 139, pp. 95–102.

13. Bakker, R. T. 1968. "The Superiority of Dinosaurs." *Discovery*, v. 3, pp. 11–22.

14. Bakker's hypothesis was disputed by his former professor, John Ostrom. Ostrom, J. H. 1985. "Social and Unsocial Behavior in Dinosaurs." *Field Museum Natural History Bulletin*, v. 55, pp. 10–21.

15. Lockley. "Dinosaur Trackways."

16. This pattern has also been inferred for large ornithopods at two other sites, where subadult-

size tracks overlap those of full-grown adults. Sarjeant, W. A. S. 1981. "In the Footsteps of the Dinosaurs." *Explorers Journal*, v. 59, pp. 164–171. The Texas herd was not progressing on a broad front like some ornithopods, however.

17. Lockley, M. G., Houck, K., and Prince, N. K. 1986 "North America's Largest Dinosaur Tracksite." *Bulletin of the Geological Society of America*, v. 97, pp. 1163–76.

18. Gore, R. 1989. "The March Towards Extinction." *National Geographic*, 175, pp. 662–99.

19. Bird, R. T. 1944. "Did Brontosaurus Ever Walk on Land?" *Natural History*, v. 53, pp. 61–7.

20. Ishigaki, S. 1989. "Footprints of Swimming Sauropods from Morocco." In Gillette, D. D., and Lockley, M. G. (Eds.) *Dinosaur Tracks and Traces*. New York: Cambridge University Press, pp. 83–6.

21. Pittman reports finding a "hind footprint that Bird did not identify." Pittman, J. G. 1990. *Dinosaur Tracks and Trackbeds in the Middle Part of the Glen Rose Formation, Western Gulf Basin, USA*. Geological Society of America Field Trip Guide no. 8, pp. 47–83.

22. Lockley, M. G., and Rice, A. 1990. "Did Brontosaurus Ever Swim out to Sea?" *Ichnos*, v.1, p. 81–90.

23. Stokes, W. L. 1957. "Pterodactyl Tracks from the Morrison Formation." *Journal of Paleontology*, v. 31, pp. 952–4.

24. Stokes, W. L., and Madsen, J. H., Jr. 1979. "Environmental Significance of Pterosaur Tracks in the Navajo Sandstone (Jurassic) Grand County, Utah." *Brigham Young University Geological Studies*, v. 26, pp. 178–84.

25. Padian, K., and Olsen, P. 1984. "The Fossil Trackway *Pteraichnus*: Not Pterosaurian, but Crocodilian." *Journal of Paleontology*, v. 58, pp. 178–84.

26. Lockley, M. G. 1990. "Tracking the Rise of Dinosaurs in Eastern Utah." *Canyon Legacy*, v. 2, pp. 2–8.

27. See Chapter 10 and Gillette, D. D., and Thomas, D. A. 1989. "Problematic Tracks and Traces of Late Albian (Early Cretaceous) Age, Clayton Lake State Park, New Mexico, USA." In Gillette, D. D., and Lockley, M. G. (Eds.) *Dinosaur Tracks and Traces*. New York: Cambridge University Press, pp. 337–42.

28. Lockley, M. G., Matsukawa, M., and Obata, I. 1989. "Dinosaur Tracks and Radial Cracks: Unusual Footprint Features." *Bulletin National Science Museum, Tokyo*, series C, v. 15, pp. 151–60.

29. Brown knew that some of his colleagues recognized the duck-bill affinity of these tracks, but he preferred to maintain a mystique and dubbed them the Mystery Dinosaur tracks. Brown, B. 1938. "The Mystery Dinosaur." *Natural History*, v. 41, pp. 190–202, 205.

30. In this popular account no details were given except for the name of the mine where the giant step had supposedly been observed. Look, A. 1955. *1000 Million Years on the Colorado Plateau*. Denver: Golden Bell, 354 pp.

31. Look, A. 1951. *In my Backyard*. Denver: Denver University Press, 316 pp.

32. The trampled flamingo comes from the Pie de Baca locality in Tepexi de Rodriguez, Mexico. The evidence suggests that an already dead flamingo was stepped on in a lakebed as a camel crossed through this ancient environment.

33. Russell, D. A., and Beland, P. "Running Dinosaurs." *Nature*, v. 264, p. 486.

34. Thulborn, R. A. 1981. "Estimated Speed of a Giant Bipedal Dinosaur." *Nature*, v. 292, pp. 273–4.

35. Kuban, G. J. 1989. "Elongate Dinosaur Tracks." In Gillette, D. D., and Lockley, M. G. (Eds.) *Dinosaur Tracks and Traces*. New York: Cambridge University Press, pp. 273–4.

Chapter 14

1. As discussed in Chapter 7, there is some doubt as to the identity of this large trackmaker. It may not have been a carnivore. Thulborn, R. A., and Wade, M. 1984. "Dinosaur Trackways in the Winton Formation (Mid-Cretaceous) of Queensland." *Memoirs of the Queensland Museum*, v. 21, pp. 243–52.

2. Leonardi, G. 1984. "Le impronte fossili de Dinosauri." In Bonaparte, J. F., Colbert, E. H. Currie, P., et al., *Solle orme de Dinosauri*. Venice: Erizzo Editrice, pp. 165–8.

3. One is beside a Kayenta Formation tracksite near Moab, Utah, the other beside a Chinle Formation tracksite near Monticello, Utah.

4. Mossman, D. J. 1990. Book review of *Dinosaur Tracks and Traces*. *Ichnos*, v. 1.

5. Look, A. 1981. *Hopi Snake Dance*. Grand Junction, Colo.: Crown Point, 64 pp.

6. Aguirrezabala, L. M., and Viera, L. I. 1980. "Icnitas de dinosaurios en Bretun (Soria)." *Munibe Société de Ciencias Nat. Aranzadi San Sebastián*, v. 32, nos. 3–4, pp. 257–79.

7. He called them Ornithoichnites, or "stony bird tracks." Hitchcock, E. 1858. *Ichnology of New England. A Report on the Sandstone of the Connecticut Valley, Especially Its Fossil Footmarks*. Boston: W. White 220 pp.

8. Tagart, E. 1846. "On Markings in the Hastings

Sands Near Hastings, Supposed to Be the Footprints of Birds." *Quarterly Journal of the Geological Society of London*, v. 2, p. 267. Jones, T. R. 1862. Correspondence. *Literary Gazette*, London, v. 8, p. 195.

9. Mudge, B. F. 1866. "Discovery of Fossil Footmarks in the Liassic Formation of Kansas." *American Journal of Science*, v. 41, pp. 174–6.

10. Mesle-Peron. 1880. "Sur de empreintes de pas d'oiseaux observées par M. deMesle dans le sud Algérie." *Comps Rendy Association France Advancement Science.*

11. Romanovsky, G. D. 1882. "The Geological Structure of the Sarvadan Brown Coal Layer in the Zervshan District." *Zapiski. SP1. Min Obschest*, ser. 2, part 17, pp. 276–96.

12. The Morrison Formation is world famous as the richest known Late Jurassic dinosaur skeletal deposit, yielding "*Brontosaurus*," *Stegosaurus*, and others from sites like Dinosaur National Monument. It was not known as a significant source of tracks until quite recently. Marsh, O. C. 1899. "Footprints of Jurassic Dinosaurs." *American Journal of Science*, v. 8, pp. 227–32.

13. Shuler, E. W. 1917. "Dinosaur Tracks in the Glen Rose Limestone, Near Glen Rose, Texas." *American Journal of Science*, v. 44, pp. 294–8.

14. Although his contribution to dinosaur tracking was small, Father Teilhard de Chardin is undoubtedly one of the most famous paleontologists because of his great renown as a theologian, author, and participant in pioneer excavations in East Asia and elsewhere. Teilhard de Chardin, P., and Young, C. C. 1929. "On Some Traces of Vertebrate Life in the Jurassic and Triassic Beds of Shansi and Shensi." *Bulletin of the Geological Society of China*, v. 8, pp. 131–3.

15. Huene, F. V. 1931. "Verschedne Mesozoische Wirbeltiereste ans Sudamerika." *Neues Jahrb. Min. Geol. Ral. Abt. B.*, v. 66, pp. 181–98.

16. Ball, L. C. 1933. "Fossil Footprints." *Queensland Government Mining Journal*, v. 34, p. 384.

17. Lockley, M. G., Houck, K., and Prince, N. K. 1986. "North America's Largest Dinosaur Tracksite: Implications for Morrison Formation Paleoecology." *Bulletin of the Geological Society of America*, v. 97, pp. 1163–76.

18. Lull, R. S. 1953. "Triassic Life of the Connecticut Valley." *Bulletin of Geology, Natural History Survey of Connecticut*, v. 81, 331 pp.

19. Casamiquela, R. M. 1964. *Estudios ichnologicos.* Buenos Aires: Collegio Industrial, Pio X., 229 pp.

20. Haubold, H. 1971. "Ichnia amphibiorum et reptiliorum fossilum." *Handbuch der Paleoher-petologie*, part 18. Stuttgart: Gustav Fischer, 124 pp.

21. Haubold, H. 1984. Wittenberg, Lutherstadt: "Saurierfahrten." Die Neue Brehm-Bucherei. A. Ziemsen, 231 pp.

22. Gillette, D. D. (Ed.) 1986. *Abstracts with Program. First International Symposium on Dinosaur Tracks and Traces.* Albuquerque: New Mexico Museum of Natural History, 33 pp. Lockley, M. G. 1986. "A Guide to Dinosaur Tracksites of the Colorado Plateau and American Southwest." *University of Colorado at Denver Geology Department Magazine*, Special Issue, no. 1, 56 pp.

23. Leonardi, G. (Ed.) 1987. *Glossary and Manual of Tetrapod Footprint Paleoichnology.* Department Nacional de Producao Mineral, Brazil, 75 pp.

24. Lockley, M. G., and Pittman, J. G. 1989. "The Megatracksite Phenomenon." *Society of Vertebrate Paleontologists*, v. 9, pp. 29–30A.

25. Gillette, D. D., and Lockley, M. G. (Eds.) 1989. *Dinosaur Tracks and Traces.* New York: Cambridge University Press, 454 pp.

26. Thulborn, R. A. 1990. *Dinosaur Tracks.* London: Chapman Hall, 410 pp.

27. Among the leading scientists who held Hitchcock's work in high regard were Richard Owen, Charles Darwin, Thomas Huxley, and Charles Lyell.

28. This conclusion, by John Ostrom of Yale, has been one of the more celebrated dinosaur discoveries in recent years. Ostrom, J. H. 1975. "On the Origin of *Archaeopteryx* and the Ancestry of Birds." *Colloquium Int. Centre Nat Rech. Sci.*, v. 218, pp. 519–32.

29. According to Ostrom and others who have studied dinosaur–bird relationships, there is no doubt about their close evolutionary ties: Birds are living dinosaurs, we can substitute the term *dinosaur* for *bird*; A dinosaur in the hand is worth two in the bush. Gautier, J. A., and Padian, K. 1989. "The Origin of Birds and the Evolution of Flight." In Padian K., and Chure, D. (Eds.) *The Age of Dinosaurs.* Short Course no. 2. Knoxville, Tenn.: Paleontological Society, pp. 121–33.

30. In "A Study in Scarlet" (1891) Conan Doyle penned the following memorable sentence: "There is no branch of detective science so important and so much neglected as the art of tracing footsteps."

31. Batory, D., and Sarjeant, W. A. S. 1989. "Sussex *Iguanodon* Footprints and the Writing of the Lost World." In Gillette, D. D., and Lockley, M. G. (Eds.) *Dinosaur Tracks and Traces.* New York: Cambridge University Press, pp. 13–18.

32. Conan Doyle, A. 1912. *The Lost World.* London: Hodder and Stroughton, 309 pp.
33. Blaiderman, C. 1986. *The Piltdown Inquest.* Prometheus, 261 pp.
34. Sternberg, C. M. 1932. "Dinosaur Tracks from Peace River, British Columbia." *National Museum of Canada Annual Report (1930),* pp. 59–85.
35. Brown, B. 1938. "The Mystery Dinosaur." *Natural History,* v. 41, pp. 190–202, 205.
36. Bird, R. T. 1985. *Bones for Barnum Brown. Adventures of a Dinosaur Hunter.* Fort Worth: Texas Christian University Press, 225 pp.
37. Bird, R. T. 1944. "Did Brontosaurus Ever Walk on Land?" *Natural History,* v. 53 pp. 61–7.
38. Farlow, J. O., and Lockley, M. G. 1989. "Roland T. Bird, Dinosaur Tracker: An Appreciation." In Gillette, D. D., and Lockley, M. G. (Eds.) *Dinosaur Tracks and Traces.* New York: Cambridge University Press, 454 pp.

Chapter 15

1. Iridium, a rare platinum-group metal that occurs in extraordinary abundance at the Cretaceous–Tertiary boundary, it is thought to be related to an event that caused widespread extinction at this time. Since tracks are evidence of live dinosaurs, documentation of tracks either above or below the Cretaceous–Tertiary boundary provides firm evidence of the activity of living animals.
2. There are many references to the "three-meter gap." David Raup devotes a whole chapter to it in his book on the extinction debate. Raup, D. M. 1986. *The Nemesis Affair.* New York: W. W. Norton, 220 pp.
3. The author discovered these tracks in 1989. A detailed report is currently in preparation.
4. Lucas, S. 1991. 'Dinosaurs and Mesozoic Chronology." *Modern Geology* (in press).

Appendix A

1. The University of Colorado at Denver Dinosaur Trackers Research Group has discovered or received reports of approximately one new site per week in the Colorado Plateau Region between 1985 and 1990.
2. Lockley, M. G. 1990. *A Field Guide to Dinosaur Ridge.* Denver: Friends of Dinosaur Ridge and University of Colorado at Denver, Dinosaur Trackers Research Group, 29 pp.
3. Several papers and booklets describe tracks from this site, though most are not in English. The most user-friendly is a nontechnical booklet published in German. Friese, H., and Klassen, H. 1979. *Die Dinosaurier Fahrten von Barkhausen im Wiehengebirge Veroffentlichungen des landkreises Osnabruk,* v. 1, 36 pp.
4. Ample literature exists on Rocky Hill and similar sites in the region, including a guide specifically on the Rocky Hill dinosaurs. Ostrom, J. 1968. *The Rocky Hill Dinosaurs.* Connecticut Geological and Natural History Survey, Guidebook no. 2, pp. 1–12.
5. James Farlow has recently written a field guide covering Dinosaur State Park and the immediate vicinity. Farlow, J. O. 1987. A guide to Lower Cretaceous dinosaur footprints and tracksites of the Paluxy River Valley, Somervell County, Texas. Fieldtrip guidebook, 21st Annual Meeting, South central section, Geological Society of America, Waco, Texas, 50 pp.
6. Several publications, including a large monograph, have been devoted to this site in the past decade. Thulborn, R. A., and Wade, M. 1984. "Dinosaur Trackways in the Winton Formation (Mid Cretaceous) of Queensland." *Memoirs of the Queensland Mus.,* v. 21, pp. 413–517.
7. The best summary of dinosaur tracksites in the Moab area is a recently published popular article. Lockley, M. G. 1990. "Tracking the Rise of Dinosaurs in Eastern Utah." *Canyon Legacy,* v. 2, pp. 2–8.
8. Hendricks, A. 1980. "Die Saurierfahrte von Münchehagen bei Rehburg-Loccum (N.W. Deutschland)." *Münster Landesmuseum Naturkunde Abhandlungen,* v. 43, pp. 3–22.
9. The Ribadessella guidebook is one of the most straightforward, colorful, and attractive produced for any dinosaur tracksite. Valenzuela, M., Garcia-Ramos, J. C., Suarez de Centi, C. 1988. *Las huellas de dinosaurios del entorno de Ribadessella.* Central Lechera Asturiana, 35 pp.
10. The Rioja guidebook is another fine example of a well conceived and colorful dinosaur footprint guide. Sanz, J. L., Moratella, J., Melero, I., and Jimenez, S. 1990. *Yacimentos paleoichnologicos de la Rioja.* Gobierno de la Rioja; Iberduero, 95 pp.
11. At present there is no guidebook to the area, though scientific publications exist both in English and in Korean. Lim, S.-K., Yang, S.-Y., and Lockley, M. G. "Large Dinosaur Footprint Assemblages from the Cretaceous Jindong Formation of Southern Korea." In Gillette, D. D., and Lockley, M. G. (Eds.) *Dinosaur Tracks and Traces.* New York: Cambridge University Press, pp. 333–6.
12. These tracks were originally described by Samuel Welles in "Dinosaur Footprints from

the Kayenta Formation of Northern Arizona." *Plateau*, v. 44, pp. 27–38. Further work is being conducted by researchers from the Museum of Northern Arizona.

13. Several publications describe the rescue operation conducted along the Peace River as tracks were rapidly excavated before they were drowned by the rising waters. Sarjeant, W. A. S. 1981. "In the Footsteps of Dinosaurs." *Explorers' Journal*, v. 59, pp. 164–71.

Index